MY GRANDFATHER'S SON

MY GRANDFATHER'S SON

A Memoir

CLARENCE THOMAS

An Imprint of HarperCollins*Publishers*
www.harpercollins.com

HarperCollins books may be purchased for educational, business, or sales promotional use. For information, please write: Special Markets Department, HarperCollins Publishers, 10 East 53rd Street, New York, NY 10022.

The author gratefully acknowledges Senator Jack Danforth for permission to reprint his letter. And Juan Williams, author and journalist, for permission to reprint part of his op-ed column.

FIRST EDITION

Designed by Joseph Rutt

Library of Congress Cataloging-in-Publication Data
is available upon request.

ISBN: 978-0-06-056555-8
ISBN-10: 0-06-056555-1

07 08 09 10 11 ID/RRD 10 9

To my mother, Leola Williams,
who gave me life;
my grandparents Myers and Christine Anderson,
who taught me how to live;
and my brother, Myers Lee Thomas,
whom I loved

CONTENTS

PREFACE

In the course of writing this book, I spent far too many solitary hours facing blank pages, digging through dusty boxes full of half-forgotten files, and plowing up long-untilled parts of my past. It was a new experience, and a strange one. I'd never been in the habit of looking back at the portion of life that I'd already lived. Most of the time, I needed all my strength to deal with the life I was then living, with all of its uncertainties, doubts, and fears. I had no idea how it would turn out. Now I know—up to a point. It is the story of an ordinary man to whom extraordinary things happened. Putting it down on paper forced me to suffer old hurts, endure old pains, and revisit old doubts. At times I was surprised by how fresh my feelings still were.

Difficult though it was to relive the past, at least I no longer had to endure the insecurity of my early years. In those days I had nothing to go on but hope, and often not nearly enough of that. I didn't know then, as I know now, that I had learned in childhood everything I needed to negotiate the challenges and difficulties that lay ahead. You never know that at the time. All you can do is put one foot in front of the other and "play the hand that you're dealt," as my grandfather so often said. That's what I did: I did my best and hoped for the best, too often fearing that I was getting the worst. In fact, though, I got everything I needed. Much of it came from two people, my grandfather and grandmother, who gave me what I needed to endure and, eventually, to prosper. They are the glue that held together the disparate pieces of my life, and holds them together to this day.

You will read about my grandparents here, along with some of the other people I met along the way. Only a few of their names will be familiar to you, and many others I have regretfully left unnamed, in the interest of brevity. Most of the people I've known, and nearly all of the important ones, are anything but famous. Yet their story is my story: their struggles in the face of futility, their perseverance through accumulated injustices, their resilience in the face of broken promises and dashed dreams, their hopefulness in the face of impenetrable bigotry, and their unrequited love for a country that often seemed to reject them at every turn.

Part of the reason for wanting to tell my story was to bear witness to what these people did for me, though I also wanted to leave behind an accurate record of my own life as I remember it. "Only the man who makes the voyage," William F. Buckley Jr., has said, "can speak truly about it." Parts of my voyage have already been written about by other people, but some of what they've written has been untrue, at times grossly so, and I didn't want to leave the telling to those with careless hands or malicious hearts. Rightly or wrongly, I have an abiding faith that my story will someday be important to someone, and so I have done my best to speak truly in these pages about my family, my friends, and myself.

I'VE ACCOMPLISHED LITTLE in my life without the help of others. This book is no different. I mostly worked on it alone, but I couldn't have done that work without God's help and the support of a long list of caring people. Always at the top of the list is Virginia, my best friend and wife, who did without my company for long stretches of writing time, endured my grumpiness as I grew weary from my protracted labors, and read each successive version of countless drafts, offering insightful editorial suggestions and boundless support. I would not have had the strength to undertake or complete this endeavor without her love.

I acknowledge with eternal gratitude Tim Duggan, my editor, who faithfully shepherded *My Grandfather's Son* into print. Terry Teachout, who trimmed the clutter of "kudzu" from my manuscript, knew almost intuitively what I was trying to say and helped me say it more clearly; in the process I lost many words, converted a manuscript into a book, and gained a friend. I also thank Sam Thernstrom, whose detailed and painstaking edits of the original manuscript were invaluable. Throughout the long process of writing, he was a calming voice of encouragement. I am grateful to Lynn Chu, my agent, who brought these three men to me, along with so many other good things.

I also acknowledge with deep appreciation the assistance of the following people who helped me at various points during the writing of *My Grandfather's Son*:

Mark Paoletta was steadfast in his support and friendship, and was always ready to do whatever was necessary to be helpful, whether digging up some obscure detail or reading drafts for accuracy. Erik Jaffe also provided invaluable assistance and advice throughout this long process. Patricia Evans at the Supreme Court Library worked tirelessly to track down even the most obscure facts and documents based on my faintest recollections. Nancy Montweiler went beyond the call of duty to assemble and provide me with all of the *Daily Labor Report*'s coverage of the Equal Employment Opportunity Commission during my tenure at that agency. Andrea Meinberg and Nicholas Matlach spent countless hours making digital copies of and restoring old photographs. Many others responded graciously when called upon to help locate documents or photos, including Jonas Jordan (photographer) of Savannah, Susan Lee of the Savannah Public Library, Michael Jordan of the Savannah History Museum (Coastal Heritage Society), Betty Purdy of the Chancery for the Diocese of Savannah, Jeff Walker of the Bethesda Home for Boys, Steve Bisson of the *Savannah Morning News*, Jewell Anderson of the Georgia Historical Society, Dr. Barbara Fertig of Armstrong Atlantic State University, Kim Cumber and Michael Southern of the State Archives of North Carolina, George Labonte of the *Worcester Telegram*, and the helpful staff of the *Donalsonville News*.

Many friends, acquaintances, and strangers have supported me with boundless supplies of prayers, kind words, and peace. I could never have continued to work for as long and as steadily as I did without the serenity I drew from them. I thank each of them from the bottom of my heart, along with family members, nuns, neighbors, teachers, and friends I've met along the way who taught me by their example or took the time to counsel me. They are too numerous to list, but I'm no less grateful for their help.

I am forever indebted to my son, Jamal, who gave me a reason to live when I thought I had none. By being mature and trustworthy beyond his years, he made it possible for me to do more than I thought I could. Jamal has always been a better son than I deserved. I have loved him since I first set eyes on him, and will do so until my last breath.

When Mark Martin, my great-nephew, came to live with Virginia and me at the age of six, we did for him what my grandparents had done for my brother and me at roughly the same age and under very similar circumstances. Raising Mark was part of what inspired me to write this book. He's been a constant source of insight into my relationship with my grandfather, and it is my hope that he benefits as much from being raised by us as I did from being raised by my grandparents.

MY GRANDFATHER'S SON

I

SUN TO SUN

I was nine years old when I met my father. His name was M. C. Thomas, and my birth certificate describes him as a "laborer." My mother divorced him in 1950 and he moved north to Philadelphia, leaving his family behind in Pinpoint, the tiny Georgia community where I was born. I saw him only twice when I was young. The first time was when my mother called her parents, with whom my brother Myers and I then lived, and told them that someone at her place wanted to see us. They called a cab and sent us to her housing-project apartment, where my father was waiting. "I am your daddy," he told us in a firm, shameless voice that carried no hint of remorse for his inexplicable absence from our lives. He said nothing about loving or missing us, and we didn't say much in return—it was as though we were meeting a total stranger—but he treated us politely enough, and even promised to send us a pair of Elgin watches with flexible bands, which were popular at the time. Though we watched the mail every day, the watches never came, and when a year or so had gone by, my grandparents bought them for us instead. My father had broken the only promise he ever made to us. After that we heard nothing more from him, not even a Christmas or birthday card. For years my brother and I would ask ourselves how a man could show no interest in his own children. I still wonder.

I saw him for the second time after I graduated from high school. He had come to see his own father in Montgomery, not far from Pinpoint, and I went there to visit him. I felt I owed it to him—he was, after all, my father, and he had let my grandparents raise me without interference—but Myers would have nothing to do with "C," as we called him, saying that the only father we had was our grandfather. That may sound harsh, but it was nothing more than the truth, for me as much as my brother. In every way that counts, I am my grandfather's son. I even called him Daddy because that was what my mother called him. (His friends called him Mike.) He was dark, strong, proud, and determined to mold me in his image. For a time I rejected what he taught me, but even then I still yearned for his approval. He was the one hero in my life. What I am is what he made me.

I am descended from the West African slaves who lived on the barrier islands and in the low country of Georgia, South Carolina, and coastal northern Florida. In Georgia my people were called Geechees; in South Carolina, Gullahs. They were isolated from the rest of the population, black and white alike, and so maintained their distinctive dialect and culture well into the twentieth century. What little remains of Geechee life is now celebrated by scholars of black folklore, but when I was a boy, "Geechee" was a derogatory term for Georgians who had profoundly Negroid features and spoke with a foreign-sounding accent similar to the dialects heard on certain Caribbean islands.

Much of my family tree is lost to me, its secrets having gone to the grave with my grandparents, but I know that Daddy's people worked on a three-thousand-acre rice plantation in Liberty County, just south of Savannah, and after their manumission they stayed nearby. The maternal side of my mother's family also came from

Liberty County, and probably worked on the same plantation, most of which has remained intact. Not long ago I saw it for the first time—during my youth blacks never went there unless they had a good reason—and found that the old barn in which my great-great-grandparents surely labored a century and a half ago is now a bed-and-breakfast inn whose Web site calls it "a perfect honeymoon hideaway." You'd never guess that slaves once worked there.

My mother, Leola, whom I called Pigeon, her family nickname, was born out of wedlock in 1929 or 1930.* Her mother died in childbirth, and she saw little of Daddy as a child. At first she was raised by her maternal grandmother, who died when she was eight or nine years old. Then she went to live in Pinpoint with Annie Green, her mother's sister. C and his family moved near there to work at Bethesda Home for Boys, which is next to Pinpoint; that was where he met Pigeon, all of whose children he sired. My sister, Emma Mae, was born in 1946, with Myers Lee following three years later. I was born between them in Sister Annie's house on June 23, 1948. I was delivered by Lula Kemp, a midwife who came from the nearby community of Sandfly. It was one of those sweltering Georgia nights when the air is so wet that you can barely draw breath. To this day my mother swears I was too stubborn to cry.

Pinpoint is a heavily wooded twenty-five-acre peninsula on Shipyard Creek, a tidal salt creek ten miles southeast of Savannah. A shady, quiet enclave full of pines, palms, live oaks, and low-hanging Spanish moss, it feels cut off from the rest of the world, and it was even more isolated in the fifties than it is today. Then as now, Pinpoint was too small to be properly called a town. No more than a hundred people lived there, most of whom were related to me in one way or another. Their lives were a daily struggle for the barest

---

* After I grew up, I started calling my mother by her first name. (Nowadays my pet name for her is "Young Lady.")

of essentials: food, clothing, and shelter. Doctors were few and far between, so when you got sick, you stayed that way, and often you died of it. The house in which I was born was a shanty with no bathroom and no electricity except for a single light in the living room. Kerosene lamps lit the rest of the house. In the wintertime we plugged up the cracks and holes in the walls with old newspapers. Water came from a nearby faucet, and we carried it through the woods in old lard buckets. They were small enough for us to fill up and tote home, where we poured the contents into the washtub or the larger kitchen buckets, out of which we drank with a dipper. We also kept a barrel at the corner of the roof to catch the rainwater in which our clothes were washed. We scrubbed them on a washboard set into the tub at an angle, rinsed them twice, wrung them by hand, and hung them out to dry.

Pinpoint was at the water's edge, and just about everyone I knew did some kind of water-related work. Many of the men raked oysters during the winter and caught crabs and fish in the spring and summer. Their boats, which we called "bateaus," could be heard far away in the marshes, straining to carry home their heavy loads. They would slowly emerge from the labyrinth of surrounding creeks and pull up to the dock, where the day's haul was unloaded. The crabs went to the crab house to be cooked, while the oysters were tossed into a bin at the oyster factory next door. Every winter the women of Pinpoint shucked oysters in the factory, a cold, damp cinderblock building whose cement floor was always wet from constant washing. (The empty building still stands by the creek, which is now clogged with weeds.) Three woodstoves stationed like sentries in the center of the room provided the only heat, and the air was heavy with the smells of burning wood, oysters, and the salt and mud of the nearby creek. The women cracked the shells and pulled out the oysters with deft, practiced movements, talking and singing spirituals as they worked. In the summer they picked crabs emptied from large baskets onto a table. The women who didn't work there served as domestics in the homes of white people, bringing back such dubious treats as the crusts of the slices of bread that had been used to make the hors d'oeuvres served at their employers' parties. When they weren't busy fishing, most of the men made extra money by doing construction, gardening, knitting fishing nets, and other odd jobs on the side.

Life in Pinpoint was uncomplicated and unforgiving, but to me it was idyllic. Myers and I skipped oyster shells on the water with our cousins and caught minnows in the creeks. We rolled old automobile tires and bicycle rims along the sandy roads, sure that there could be no better fun. We had only two store-bought toys, a beat-up old red wagon and a little fire truck that I could pedal. Instead we played with "trains" made out of empty juice cans strung together with old coat hangers and weighted with sand, and we also made gunlike toys called "pluffers" out of canes and wooden mop or broom handles, using green chinaberries as ammunition. We were supposed to stick close to home, but no sooner did the adults leave for work each day than we ran for the sandy marshes, in which we hunted for fiddler crabs. We wandered through the nearby woods, sometimes tussling with the white kids from Bethesda Home for Boys, the oldest orphanage in America, an oasis of big brick buildings and expansive, well-kept lawns. Any attempt to invade this paradise was fiercely resisted by the lucky boys who lived there.

Nothing about my childhood seemed unusual to me at the time. I had no idea that any other life was possible, at least for me. Sometimes I heard the grown-ups talk about the white people for whom they worked, but I took it for granted that they were all rich. Photographs in newspapers and magazines gave me fleeting glimpses of an unreal existence far from home, but Pinpoint was my world, and until I started going to school, the only sign that there might be another one was the occasional airplane or blimp I saw flying overhead.

I entered Haven Home School, the local grammar school for blacks, in the fall of 1954. I walked with my sister and the older kids to the end of Pinpoint Road to wait for the school bus, and I could barely contain my excitement when it came rolling around the bend at last. That was the year when Myers and Little Richard, our cousin, burned down Sister Annie's ramshackle house, in which we still lived, while playing with the matches used to light the stove and lamps. As I came home from the school bus, I saw that the house had become a smoldering pile of ashes and twisted tin. After that, Pigeon took my brother and me to Savannah, where she was keeping house for a man who drove a potato-chip delivery truck. We moved into her one-room apartment on the second floor of a tenement on the west side of town. Emma Mae stayed with Annie, who went to live with one of our cousins until she could build another house.

Nowadays most people know Savannah from reading *Midnight in the Garden of Good and Evil*. To them it is an architectural wonderland full of well-heeled eccentrics and beautifully preserved eighteenth- and nineteenth-century houses. I didn't live in that part of town. When I was a boy, Savannah was hell. Overnight I moved from the comparative safety and cleanliness of rural poverty to the foulest kind of urban squalor. The only running water in our building was downstairs in the kitchen, where several layers of old linoleum were all that separated us from the ground. The toilet was outdoors in the muddy backyard. The metal bowl was cracked and rusty and the wooden seat was rotten. I'll never forget the sickening stench of the raw sewage that seeped and sometimes poured from the broken sewer line. Pigeon preferred to use a chamber pot, and one of my Saturday-morning chores was to take it outside and empty it into the toilet. One day I tripped and tumbled all the way down the stairs, landing in a heap at the bottom. The brimming pot followed, drenching me in stale urine.

Pigeon enrolled me in the afternoon first-grade class at Florance Street School, one of the first public schools in Savannah built specifically for black students. Myers wasn't old enough to go, so she made other arrangements for him. I don't know where he went, but she took him with her early each morning, leaving me to fend for myself until it was time for me to walk to school. Breakfast, when I had it, was usually cornflakes moistened with a mixture of water and sweetened condensed milk. (We couldn't afford sugar.) Pigeon always came home tired and drained. It was as though her job sapped all the hope out of her. She worked to stay alive and keep us alive, nothing more.

Many of the kids I had known at Haven Home School were relatives of one sort or another, but now I was alone, a stranger in an ugly, unfamiliar world. My lessons were slow moving and repetitive, so I started skipping school and wandering around the west side of town, mystified and curious. In the evenings and on rainy days, I would spend hours sitting by our solitary second-story window, watching the parade of people walking up and down the dirt street that ran past our building. Men gathered around open fires, talking as they warmed themselves. Sometimes shrill-sounding sirens pierced the night. Their wail might mean an arrest, an accident, or a fire, all of them welcome breaks in the bleak monotony of my new life.

It didn't help that our apartment was so uncomfortable. My mother and brother shared the only bed, leaving me to sleep on a chair. It was too small, even for a six-year-old. We couldn't afford to light the kerosene stove very often, so I was cold most of the time, cold and hungry. Though there was only one store in Pinpoint, the rivers and the land had provided us with a lavish and steady supply of fresh food: fish, shrimp, crab, conch, oysters, turtles, chitterlings, pig's feet, ham hocks, and plenty of fresh vegetables. Never before had I known the nagging, chronic hunger that plagued me in Savannah. Hunger without the prospect of eating and cold without the prospect of warmth—that's how I remember the winter of 1955.

Late that summer we moved to a two-bedroom apartment on East Henry Lane. Our new neighborhood was just as poor, but a better place to live. We had a kitchen with a stove and refrigerator. The outdoor toilet didn't leak. The beds were old, but at least I had one of my own, and it was big enough for a growing boy. My mother's father lived nearby on East Thirty-second Street, and we visited him and our grandmother several times that summer. Their house, which had been built three years earlier out of cinder blocks, was painted a gleaming white. It had hardwood floors, handsome furniture, and an indoor bathroom, and we knew better than to touch anything.

One Saturday morning in August, Pigeon told Myers and me that we were going to live with our grandparents. Without a word of explanation, she dumped our belongings into two grocery bags and sent us out the front door. We walked straight to our grandparents' house, passing by the places that were to become the landmarks of our childhood: Mr. Lee's grocery store, Mr. Goodman's grocery store on the opposite corner, Mr. Moon's fish market, Reverend Bailey's shoe store, the Polar Bear Ice and Coal Company, and the Meddin Brothers meat-packing company. It was only two and a half blocks from East Henry Lane to the spotless white house on East Thirty-Second Street, but in all my life I've never made a longer journey.

I don't know the whole story of Pigeon's decision to send Myers and me to live with our grandparents. The main reason, though, must have been that she simply couldn't take care of two energetic young boys while holding down a full-time job that paid only ten dollars a week. C did nothing—and I mean nothing—to help us. A court had ordered him to pay child support to Pigeon, but he ignored it. Since she refused to go on welfare, she needed some kind of help, and I suspect my grandfather told her that we would either live with him permanently or not at all. He didn't like complicated arrangements, and he may also have feared that giving us up after a short time would be too hard for him to bear. He often told us about a dog he had once owned that had strayed from the yard and was run over by a car. My grandfather was so devastated by the dog's death that he never owned another one after that.

My grandparents were in their forties when they took us in, and had been married for more than two decades without having any children together. (Daddy had had two children out of wedlock before he married my grandmother, the second of whom was Pigeon, and my grandmother bore an illegitimate child of her own who died in infancy.) Perhaps they hoped that Myers and I might fill the empty place in their home and their hearts.

MY GRANDMOTHER'S NAME was Christine, but we always called her Aunt Tina (pronounced "Teenie"), as our cousins did, and she never asked us to call her anything else. Aunt Tina was a slight woman, five foot two and thin, with perfect teeth. She was soft spoken and easygoing, and I thought she was beautiful. Her favorite book was *To Kill a Mockingbird*, Harper Lee's novel about a black man falsely accused of raping a white woman. She must have bought it on one of her train trips to visit her sister, who lived in the Bronx. I doubt she ever read it all the way through—she had only a sixth-grade education—but she knew the story well, and in 1962 she took Myers and me to see the movie version. Aunt Tina must have been shocked to see that all our possessions fit into one grocery bag apiece, but she never said so. She showed us to our bright, airy new room, which contained matching twin beds with frilly white bedspreads, a chest of drawers, and a closet of our very own, and told us we were home now. From then on she did all she could to make that home warm and loving.

Myers Anderson, my grandfather, stood five foot eleven and weighed 195 pounds. He had proud West African features—high cheekbones, a broad nose, and very dark skin—and a thin mustache that he trimmed each morning with a Gillette razor. Daddy had only nine fingers, having lost his right index finger working on a boat. His hard, wiry body was strongly muscled, his deep bass voice intimidating. He had only a third-grade education, which he said amounted to nine months of actual learning since he went to school for only three months a year. It would be too generous to call him semiliterate; his pride made him blame his reading difficulties on bad eyesight, but the truth was that he could barely read at all. By the time I was in third grade, I was the best reader in the house. As long as he lived, though, Daddy struggled mightily with the newspaper and the Bible, and once he mastered a passage of Scripture he would read it over and over again.

Daddy rarely spoke of his own father except to say that he'd been a "jackleg preacher," meaning that he was self-taught. Perhaps he didn't know much more than that, and preferred to keep what he did know to himself. I do know, however, that Daddy was illegitimate, and that he resented his father's lack of interest in him. He believed that he had been born in Savannah in 1907, and later on he chose April 1 as his birthday, in the same way that so many other southern blacks picked out similar birthdays for themselves. (Louis Armstrong chose July 4, 1900.) His mother died when he was nine, and he went to live with his maternal grandmother, Annie Allen, in the Sunbury area of Liberty County. Annie died when Daddy was twelve, and he was turned over to her son and his wife, who lived nearby. Uncle Charles and Aunt Rosa already had thirteen children of their own, so he was just another mouth to feed. He often told us that he had spent his whole childhood being "handed from pillar to post." Daddy always spoke of his grandmother, a freed slave, with reverence. A no-nonsense woman who didn't spare the rod, she used to spit on the side of the road on a hot summer day and warn Daddy that he had better run his errand before the spittle dried if he wanted to escape a whipping; Uncle Charles beat him, too, sometimes with an oar, knocking him off his boat. Yet in spite of their harsh treatment—or because of it—Daddy knew they'd cared enough to raise him right, and he felt he owed them a debt of gratitude too great to repay. When Uncle Charles died in 1960, Daddy, Myers, and I knelt by the bed where his body lay and said the rosary.

In his youth Daddy worked on Uncle Charles's boat, made bootleg liquor, and moved houses, among other things. Later on he bought an old truck and started cutting wood at night and selling it during the day. He soon added coal and ice delivery to his services, followed by fuel oil. By the summer of 1955, he had a well-established family business that he and Aunt Tina ran out of their home. A frugal, industrious man, he also owned several rental houses in the neighborhood, and he made the cinder blocks used to build two of the rental houses and the house in which we lived. Our block was an island of safety tucked in between two much rougher blocks. Bootleg liquor, known in Savannah as "scrapine," "shine," or "scald," was sold on the 600 block of East Thirty-second Street, where there were also gambling and other illegal activities. Daddy called it Blood Bucket, and we often heard police cars and ambulances racing to and from that neighborhood in the middle of the night. We were strictly forbidden to set foot in Blood Bucket, which was on the other side of the railroad tracks that ran by our house, but we often saw drunks (some of whom were relatives of ours) roving up and down our own block, often in packs, looking for a drink of scrapine "'cross the track." Myers and I had to be careful not to bump into them as we walked to and from school. One morning we had to step over the body of a man passed out on the sidewalk. Vomit and spittle flowed from the corner of his mouth, forming a puddle near his head. Daddy often spoke of the man later, both because our childish disgust amused him and in order to remind us of the dangers of drunkenness. A moderate, self-disciplined man who took only one drink each evening, he said that black people had enough trouble getting through life without adding liquor to their list of woes. "Why make matters worse?" he asked. It was one of many lessons he would teach us.

"The damn vacation is over," Daddy had told us on the morning we moved into his house. I thought of the filthy outdoor toilet behind Pigeon's old tenement and tried to figure out what vacation he was talking about. He went on to explain that while our mother had allowed us to come and go as we pleased, there would be "manners and behavior" and "rules and regulations" from now on. The first rule was that Aunt Tina was always right—and so was Daddy. Whenever we failed to obey either of them, punishment was swift, sure, and painful. Daddy didn't whip us regularly, but our encounters with his belt or a switch were far from infrequent, and it soon became clear that he meant to control every aspect of our lives: there would be no more carefree days spent wandering through the marshes hunting for fiddler crabs, no more roaming the streets instead of going to school. For a time we wondered why our real father didn't come and rescue us, but we had long since accepted our fate by the time we finally met him.

In return for submitting to Daddy's iron will, Myers and I lived a life of luxury, at least by comparison with Sister Annie's Pinpoint shanty or the apartments we had shared with Pigeon. The home that Daddy and Aunt Tina built for themselves (and in which my mother now lives) had two bedrooms, one bathroom, separate living and dining rooms, a den, and a kitchen. My grandfather often boasted it only cost six hundred dollars. It wasn't all that large, but to us it looked as big as a palace. We had our own beds, plenty to eat, and an endless supply of soft drinks and fresh milk. The appliances—a Magic Chef stove, a refrigerator, a separate freezer, a hot-water heater, and a Kenmore washer—were all new, and new to us, too. I had never before seen a house with such conveniences, or with an indoor porcelain toilet that worked. I flushed it as often as I could in my first months on East Thirty-second Street.

Daddy made it plain, though, that there was a connection between what he provided for us and what he required of us. He told us that if we learned how to work, we would be able to live as well as he and Aunt Tina did when we grew up. That, he said, would be our inheritance. For him all honest work was good work, and he proudly wore a T-shirt and khaki work pants (or, during cooler weather, a khaki shirt or sweatshirt) every day, but he expected us to do better. Our first task was to get a good education so that we could hold down a "coat-and-tie job," and he wouldn't listen to any excuses for failure. "Old Man Can't is dead—I helped bury him," he said time and again. It wasn't easy for us to accept his unbending rules, but we did it anyway; he gave us no choice. The door to his house, he said, swung both ways. It had swung inward on our arrival, but if we didn't behave, he warned ominously, it would swing outward. He added that he would never tell us to do as he said, but to do as he did—and he kept his word.

VIRTUALLY EVERYONE IN my family, including Aunt Tina, was either a Baptist or a member of one of the "sanctified" Holiness churches to which so many southern blacks belonged, but Daddy had gone his own way and converted to Roman Catholicism in 1949. The orderly Catholic liturgy and the discipline of the nuns and priests were more to his liking than the endless Baptist services and the overemotional preachers who led them, and he also thought that children learned better when they wore uniforms and studied in a structured environment. It followed that Catholic schools had to be better than public schools, so he sent my brother and me to one.

I spent the first day of my second-grade year at East Broad Street School, a public elementary school in which Daddy briefly enrolled me while he tried to get me into St. Benedict the Moor Grammar School in spite of my dismal attendance record at Florance Street School. It took him just one day to win his case. This was in September 1955, and even though the Supreme Court had ruled the year before that segregation was unconstitutional, the public schools of Georgia had chosen to defy it. I knew nothing of *Brown* v. *Board of Education*, of course. I was too young to understand such things. The only responses to the decision that I can recall from my early days in Savannah were the "Impeach Earl Warren" billboards that started cropping up here and there and the adoption in 1956 of a new state flag that bore the seal of Georgia next to the Confederate flag. In any case I was far more interested in going downtown for the very first time to shop for my school uniform: white shirt, blue tie, and blue slacks.

St. Benedict's was run by the Missionary Franciscan Sisters of the Immaculate Conception, most of whom were Irish immigrants. The building was old and in need of repair, but it was clean and neat. The students helped keep it that way. Since there was no maintenance staff other than Mr. Newton, who also took care of the parish church and St. Pius X, the Catholic high school for blacks, we were required to pick up trash, empty wastebaskets, sweep floors, and clean blackboards. The nuns were far more demanding than the teachers at Florance Street School. They expected our full attention and made sure they got it, dispensing corporal punishment whenever they saw fit. Classes were large—around forty students—but orderly. We began the day's studies with the catechism and said the rosary in class each afternoon. We learned that God made us to know, love, and serve Him in this world, and to be happy with Him in the next. The sisters also taught us that God made all men equal, that blacks were inherently equal to whites, and that segregation was morally wrong. This led some people to call them "the nigger sisters."

A few of my classmates were the children of teachers, doctors, and small businessmen. Most, though, came from the same kind of environment in which I had grown up, and many were worse off than I had been. The annual tuition was only $25, but that was a lot of money for a black family in Georgia in 1955. Some parents couldn't afford to buy their children new uniforms, so they wore old ones that may have been given to them. Whatever our circumstances, the nuns treated us all with respect and insisted that we do our best, though some of them insisted harder than others. Sister Mary Virgilius, my eighth-grade teacher, took a no-nonsense, high-expectations approach to teaching, which may be why she got the most out of me. She knew I was an underachiever who was able to get good grades without putting in enough effort to realize my full potential, and when I got an unusually high score on my high school entrance exam, she called me on it. "You lazy thing, you!" she said.

Every morning before school, Aunt Tina fed Myers and me bacon, sausage, eggs, and grits, except on Friday, the Catholic day of fasting, when we ate fish instead. Afterward she fixed bag lunches for us to take to school, and Daddy left milk money on the table. It took fifteen or twenty minutes to walk to St. Benedict's, but we were never assaulted or threatened, and we missed school only once in all the years I lived on East Thirty-second Street. That was not Daddy's fault: he warned us that if we died, he'd take our bodies to school for three days to make sure we weren't faking, and we figured he meant it. He also told us that our teachers, like Aunt Tina, were always right. Even when they weren't, it did no good to complain to him. Doing so was sure to get us in worse trouble.

I loved St. Benedict's, in part because the nuns, for all their strictness, seemed almost lenient by comparison with my all-seeing grandfather. I reveled in every opportunity they gave us to leave the classroom, whether it was to walk to the Catholic cathedral for a special event or go to the train yard for an evacuation drill. Best of all was the annual school trip to a black beach in Hilton Head, South Carolina, though I didn't enjoy these outings nearly as much after I almost drowned. Not knowing how to swim, I'd always been careful to stay close to shore, but this time a strong current sucked me out into the ocean, and I panicked and went under. Then I miraculously bumped into someone who pulled me up and brought me back to shore, coughing all the way. As far as I was concerned, it was St. Michael the Archangel, my patron saint, who had saved me.

School let out at two-thirty, and Myers and I were expected home by three o'clock sharp. Aunt Tina would give us a glass of milk and a snack, and if Daddy didn't have any work for us to do, we went back outside to ride our bikes, skate, or play games with the neighborhood kids. Daddy had put up a basketball goal in the backyard when we'd first come to live with him, but when he saw older boys using it, he took it back down and stored it in the garage. After that we made our own makeshift goals from galvanized-steel trash cans whose bottoms were rotted out, which we nailed to a pole. We rarely had a backboard, and our basketballs were leaky and lopsided. This made it hard to dribble, but that didn't matter, since the dirt lane on which we played was covered with rocks, glass, nails, and other debris. We also played one-handed touch football, "half rubber" (similar to stickball), and a game of our own devising that we called "send back." We almost always played in the street or the lane behind our house, meaning that we had to watch out for cars, trees, and poles. In between games we tinkered with our bikes, adjusting the chains and sprockets or patching the inner tubes. We also played with hula hoops and yo-yos, shot marbles, and played hopscotch and jacks. When it rained we made paddleboats to sail in the puddles.

It was around this time that I started reading comic books. My favorites were *Rawhide Kid, Two-Gun Kid,* and *Kid Colt, Outlaw,* the misunderstood western heroes, but I also liked superhero comics, particularly *Superman, Green Lantern, Flash,* and *Spiderman.* In addition Myers and I were allowed to watch a limited amount of television, always asking permission before turning on the set. (Daddy rarely said yes, so we asked Aunt Tina whenever we could.) We liked *Hopalong Cassidy, The Roy Rogers Show, The Cisco Kid, The Adventures of Superman, Sky King,* and any other series that was full of action. On special occasions Aunt Tina let us stay up past our nine o'clock bedtime to watch *The Twilight Zone, Alfred Hitchcock Presents,* or *Have Gun Will Travel.*

Except for the evening news, Daddy didn't have much use for TV, though he sometimes watched baseball on Saturday afternoons in the spring, especially when Pee Wee Reese and Dizzy Dean were calling the games. He played checkers with friends in the lane behind our house and took part in regular games of Pokeno and bid whist when we'd first come to live with him, but came to frown on most leisure-time activities as Myers and I grew older. He wouldn't let us play on sports teams or join the Cub Scouts, though visits to the nearby Carnegie Library were always allowed, since the librarians, like the nuns of St. Benedict's, helped us learn. I spent countless hours immersed in the seafaring adventures of Captain Horatio Hornblower, the gridiron exploits of Crazy Legs McBain, and the real-life triumphs of Bob Hayes, the world's fastest man; I also read about the civil-rights movement, of which I still knew next to nothing. I was never prouder than when I got my first library card, though the day when I'd checked out enough books to fill it up came close.

No matter what we were doing, playtime was over at the sound of the five o'clock whistle, which blew at the laundry down the street.

That meant it was time for dinner, which we ate together at the porcelain-topped kitchen table. Myers sat across from Aunt Tina, putting Daddy in an ideal position to whack him on his left hand whenever he tried to eat with it. Until then Myers had been left-handed, but Daddy, like so many people of his generation, was sure that there was something wrong with being left-handed, and his word, as always, was law. We ate fried pork chops, stew beef, deer, raccoon, or chicken, accompanied by rice and vegetables—usually collard greens, butter beans, or snap beans—simmered for an hour or two with neck bones or fatback. On Fridays we ate fish. Daddy liked it stewed and served over grits, and he always ate the head. He lectured us sternly about our refusal to do the same, saying that we had it too good. "Hard times make monkey eat cayenne pepper," he added, then told us how his grandmother used to eat cow's eyes. This made us shiver with horror, but Aunt Tina took pity on us, and from then on she fried our fish with the heads off.

After dinner my grandparents watched the news and then, if it wasn't too cold, sat on the front porch, smoking cigarettes and chatting while Myers and I washed the dishes and swept out the kitchen. If it was still daylight when we were done, we went back outside to play. Otherwise we did our homework or watched TV, though Daddy went to bed at eight o'clock, after which we had to keep as quiet as possible. He insisted that we bathe in what he called a "teaspoon" of water, using laundry detergent instead of soap. "Waste not, want not," he repeatedly warned us. We weren't allowed to use towels to dry ourselves, either, since Daddy thought our washcloths were good enough to get us dry (as well as being easier to launder than towels). Whenever he thought we hadn't gotten ourselves clean enough, he finished the job himself, a terrifying experience that we did everything we could to avoid.

On Saturdays we got up before sunrise, usually finishing our weekend breakfast of cereal and milk as Daddy came home from his weekly haircut. If we overslept, he would poke his head into our room and shout, "Y'all think y'all rich." More often than not, though, we were glad to get up early, since we preferred cold cereal to grits, especially when we could sneak a little extra sugar from the bowl. Our own hair was trimmed every other week, though it was cropped so close that "shaved" might be a better word. The first barber I remember was a preacher who kept a shop on East Broad Street. The reverend would place a board across the arms of the barber chair, hoist us up into this makeshift seat, and go about his rough business, pressing his clippers hard against our scalps, shearing the sides and back clean, and leaving no more than a trace of hair on top. That was how Daddy wore his hair and how he made us wear ours, and he didn't care that our friends teased us for looking unstylish.

A steady stream of mysterious-looking people passed through the dimly lit shop. We were particularly interested in a man in a long overcoat who dropped by for hushed and hurried conversations with the reverend, followed by an exchange of money and a hasty departure. Daddy told us that he was the "bolita man," the local numbers runner, and smiled at our wide-eyed reaction. As we waited to get our hair cut, we overheard the other customers' noisy discussions about sports and politics, and flipped through the raggedy old magazines and newspapers that the reverend left for his patrons to read. It was in his shop that I saw my first copies of *Ebony*, *Jet*, and the *Pittsburgh Courier*, all of which catered to black readers.

We went to church every Sunday, and I started studying to be an altar boy in the third grade. This was before Vatican II permitted Catholic priests to say Mass in English, meaning that we had to learn the Latin responses by heart. I soon began serving the six a.m. Mass in church and, from time to time, the Mass at the convent next door. Daddy liked to watch us serve Mass and insisted that we take our responsibilities seriously, including the important task of being painstakingly dressed and groomed. One Sunday he saw that my shoes were unpolished and warned me that I couldn't serve Mass in them. I replied that if I stopped to polish them, I would be late for church. The next thing I knew, I was lying in a crumpled heap against the wall with Daddy standing over me saying, "You better not be late for Mass, boy!" I had broken another of his cardinal rules—we were never to "'spute" his word—and so he slapped me across the face. I polished my shoes as fast as I could and ran all the way to church, getting there just in the nick of time.

Myers and I were allowed to run and play almost at will during our first two summers at East Thirty-second Street. We went barefoot, as we had done in Pinpoint, and once I stepped on a rusty nail that sank deep into my foot and had to be pulled out. My grandmother placed a penny and a piece of bacon on the wound and tied them in place with a strip of white cloth, and soon I was playing again—with shoes on. Aunt Tina had any number of home remedies for what ailed us. In the winter we took weekly doses of cod-liver oil to ward off colds. We were so small when we came to live with her that she correctly assumed we had worms, which she sought to clean out of our systems by giving us regular doses of castor oil and (on occasion) a mixture of sugar and turpentine, both of which seemed far worse than any diseases they could possibly have cured.

Daddy and Aunt Tina occasionally took us to drive-in movies in the summer, and on holidays we would ride out to the country in their 1951 Pontiac to visit our relatives. I especially liked Cousin Hattie Chip and her husband, Cousin Robert, whose chicken yard was full of plum and fig trees. I loved to pick and eat the fruit, but the yard was full of chicken droppings, and since I was barefoot more often than not, I had to watch where I stepped. I got myself in a jam when I decided to knock down a wasp nest near the top of the roof of their house. I threw rocks at it until I scored a direct hit, and all at once the air was full of furious wasps chasing me all around Cousin Hattie's yard, delivering their painful retribution. I got no sympathy from Daddy, who only asked whether I'd learned my lesson.

The carefree days ended when I entered fourth grade and became old enough to help with the family business. Daddy had stopped selling wood and coal by the time Myers and I went to live with him; most of Savannah's residents had switched by then to fuel oil or natural gas. "Fuel oil is our bread and butter," he told us. Our telephone number, ADam 3–2303, was painted on the sides of both of his trucks, and Aunt Tina, who had worked as a maid in her younger years, now stayed home to keep house and take orders over the phone, which rang constantly in the late fall and winter. Daddy got up between two-thirty and four each morning to start making his rounds, and I worked every day after school and all day on Saturdays, alternating between taking orders and riding in the truck with him.

A year after Myers and I went to live with Daddy, he bought a brand-new GMC truck. He took out the heater, explaining that it made no sense to keep the cab warm when we had to get out at every stop. The warmth, he said, would only make us lazy, and he also thought we would catch colds from constantly getting in and out of a warm cab. We weren't allowed to wear gloves when pulling the hose or putting oil in the tanks—he was afraid they might get caught in the machinery—and my fingers grew numb from the cold. Sometimes when Daddy left the truck to make a phone call or go to the bathroom, I took some of the old newspapers that we kept in the truck to clean up spills and set them on fire to thaw out my hands.

IN THE FIFTIES and sixties, blacks steered clear of many parts of Savannah, which clung fiercely to racial segregation for as long as it could. The Ku Klux Klan held a convention there in 1960, and 250 of its white-robed members paraded down the city's main street one Saturday afternoon. No matter how curious you might be about the way white people lived, you didn't go where you didn't belong. That was a recipe for jail, or worse. That's why I never saw much of the fancy houses in the center of town, except through the windows of Daddy's truck. Vast though it seemed to me, the part of Savannah in which we lived was almost as small as Pinpoint. Yet it offered plenty of chances for an idle boy to misbehave, and Daddy had every intention of keeping us out of trouble, even if he had to build a new house in order to do it.

On Christmas Day, 1957, we drove out to Liberty County to spend the holiday with Cousin Hattie and Cousin Robert. That afternoon Daddy took Myers and me for a ride. We went to the place where he had grown up, which we called "the farm" even though the fields had been left untended for years and now lay fallow. He stopped the car and got out, and we followed him to a spot in the shadow of several large oak trees. The sixty acres on which we stood belonged to the Allen family. They had been passed down undivided from generation to generation, as was often customary with land owned by southern blacks. Any family member was entitled to live there, but most of Daddy's relatives had moved up north or to the city by then. He looked around in silence. Then he told us that we were going to build a house on that very spot.

By springtime we'd finished building a simple four-room house, and we spent the summer building garages, a barn, and other facilities, putting up fences, and clearing the surrounding land with axes and bush hooks. Friends and family members had helped us lay the cinder blocks and put on the roof, but we did all the rest of the work ourselves, screening the porches and installing a secondhand tub, sink, and toilet Daddy had found at a place that looked more like a junkyard than a store. From then until 1967 I lived there each summer, going out to the farm on the last day of school and returning to town the night before the fall term started. Years later Daddy explained that he'd decided to build a house and cultivate the family land in order to keep Myers and me off the streets of Savannah during the hot-weather months when nobody bought fuel oil. At the time, though, he told us nothing. He and Aunt Tina spent their last days in that little house. For me it was a place of torment—and salvation.

Now that I was ten years old, Daddy expected me to pull my load on the farm. No sooner was one task finished than we went straight on to the next one. We got up before sunrise and started in at once, pausing for breakfast around eight o'clock. We stopped again at noon for lunch, after which Daddy napped until two. This gave Myers and me some free time, though we were expected to be ready to resume work as soon as Daddy woke up. We skipped the midday break whenever we were out in the woods cutting trees, clearing land, or working on the fence. On those days we went back to the house at noon to pick up the lunch Aunt Tina had made for us, then carried it back out to the woods to eat. The workday was over at five or six, but all that meant was that we took up such lesser chores as cutting grass, washing dishes, and feeding the animals.

Daddy borrowed Cousin Jack Fuller's horse to plow the fields during our first few years on the farm. (Later on he bought a secondhand tractor that he taught me to drive.) Lizzie was a spirited animal with a crazed look in her eye, and she walked fast when pulling a plow, dragging Daddy behind her as we chased him down the freshly plowed furrows. Once she pulled him along so fast the plow handle broke several of his ribs. We usually went barefoot, and the cool, freshly plowed soil was a relief to our feet as the temperature rose each morning, though we sometimes had scary encounters with snakes unearthed by the plow. They sent us scurrying for a few breathless moments, but we always returned as quickly as possible to the task of planting, spreading fertilizer, or straightening the plants that had been knocked over by the horse or the plow. We knew we had to finish before the sun rose so high that it grew too hot to work. When the plowing was finished, we weeded the field with hoes, going from plant to plant, row by row. It was a slow, tedious job, though easier than picking beans, which required us to sort each pod in order to determine whether it was ripe. Except in the early morning hours, the Georgia sun beat down on us brutally and unrelentingly. The temperature routinely exceeded a hundred degrees by summer's end, and even the cooler morning air was full of gnats, mosquitoes, and flies. The house had a tin roof that kept out the rain but radiated the heat, and the only breeze came from a single window fan in my grandparents' bedroom. My pillow would be soaked with sweat by the time I finally fell asleep.

We picked the corn that we ate, along with the other vegetables that we grew, but the corn that fed our hogs during the winter was left in the field until fall. We came back to the farm in September to harvest it, a chore that kept us too busy to take part in such after-school activities as team sports. In late August, when the sun was at its highest and hottest, we cut sugarcane. Sugarcane leaves are covered with sharp fibers that stick to your skin and are all but impossible to remove, so we had to wear long-sleeved shirts to harvest the cane, which made us even hotter. Once the cane was cut, we placed it in an earthen bank, where it remained until it was time to make syrup in the winter, though Daddy always put aside a few stalks for us to eat. They were wonderfully sweet and tasty.

Whenever Daddy shot a deer, raccoon, or squirrel, we had to skin and clean it for him, just as we cleaned the fish he caught late at night. He would carry home a couple of washtubs full of mullet around four or five in the morning, waking us up to announce his success. Then he went to bed, leaving Myers and me to clean his catch at an outdoor bench lit by a hundred-watt bulb whose light seemed to draw every insect in southeast Georgia. It took us hours to finish the dirty job, and I spent them longing to be clean and dry again and to escape the gnats and mosquitoes that swarmed around us. We also raised chickens, killing them one at a time for meals. Before returning to Savannah at the end of the summer, Myers and I caught the remaining birds, wrung their necks, dipped them in hot water, plucked the feathers, and cleaned them, setting aside the gizzards and liver. The rest we took back to town to freeze and eat through the winter. The feet were a special delicacy that Aunt Tina liked to stew and serve with gravy and rice.

From time to time we slaughtered one of the forty or so hogs we kept. Daddy would shoot it in the head with his .22 rifle, then cut the jugular vein to bleed out the carcass. We then placed it in a fifty-five-gallon barrel half full of water, set into the ground at an angle and surrounded by a fire. We slid the hog in and out of the barrel, scraping its skin to remove the coarse hair. Then we cut a slit between the bone and tendon of the hind legs, slid in a short, strong stick, and lifted the whole hog onto a rack hung from a tree branch, six feet or so above the ground. Daddy cut the hog open from tail to head, and its guts fell into a tub placed underneath the carcass. We saved nearly every part of the animal, making fresh crackling from the skin and using the intestines for chitterlings. Portions were given to friends and relatives, while the rest went into the freezer to be saved for a rainy day. Daddy always seemed to be preparing for rainy days. Maybe that's why they never came.

Our small, soft hands blistered quickly at the start of each summer, but Daddy never let us wear work gloves, which he considered a sign of weakness. After a few weeks of constant work, the bloody blisters gave way to hard-earned calluses that protected us from pain. Long after the fact, it occurred to me that this was a metaphor for life—blisters come before calluses, vulnerability before maturity—but not even the thickest of skins could have spared us the lash of Daddy's tongue. "I could do more with a teaspoon than you can do with a shovel," he snapped whenever we were shoveling dirt. "You worth less than a carload of dead men." He never praised us, just as he never hugged us. Whenever my grandmother urged him to tell us that we had done a good job, he replied, "That's their responsibility. Any job worth doing is worth doing right."

What bothered us most about life on the farm, though, was that it kept us from spending our summer vacations playing with our friends in Savannah. That was what Daddy had in mind. He feared the evil consequences of idleness, and so made sure that we were too busy to suffer them. In his presence there was no play, no fun, and little laughter. "No time for that kind of foolishness," he would say. Genesis 3:17–19 was his credo: "Cursed is the ground because of you; through painful toil you will eat of it all your life. It will produce thorns and thistles for you, and you will eat the plants of the field. By the sweat of your brow you will eat your food until you return to the ground, since from it you were taken." Because of man's fall from the Garden of Eden, Daddy said, it was our lot in life to work "from sun to sun." Once, years later, I got up the nerve to tell him that slavery was over. "Not in *my* house," he replied.

THE FAMILY FARM and our unheated oil truck became my most important classrooms, the schools in which Daddy passed on the wisdom he had acquired in the course of a long life as an ill-educated, modestly successful black man in the Deep South. Despite the hardships he had faced, there was no bitterness or self-pity in his heart. As for bad luck, he didn't believe in it. Instead he put his faith in his own unaided effort—the one factor in life that he could control—and he taught Myers and me to do the same. Unable to do anything about the racial bigotry and lack of education that had narrowed his own horizons, he put his hope for the future in "my two boys," as he always called us. "I am going to send you boys to school and teach you how to work so you can have a better chance than I did," he said. We were his second chance to live, to take part in America's opportunities, and he was willing to sacrifice his own comfort so that they would be fully open to us.

Even then I understood that he had rescued me from difficult circumstances, but it was not until long afterward that I grasped how profoundly Daddy, Aunt Tina, and the nuns of St. Benedict's had changed my life. Sometimes their strict rules chafed, but they also gave me a feeling of security, and above all they opened doors of opportunity leading to a path that took me far from the cramped world into which I had been born. In Pinpoint I was a little Negro boy growing up among hardworking but uneducated people. From there I moved to the confusion and squalor of a run-down tenement in Savannah, where I led a life of being cold and not knowing when I would feel warmth again, of constant, gnawing hunger and not knowing when I would eat again, a life in which knowledge trickled in by the thimbleful when I yearned for floods of truth. To stay there would have doomed me to a dismal life of ignorance, perhaps even of crime—a life lost before it started.

I know all this now, but back then Myers and I were still young children, full of spunk and unused to the yoke of discipline, and we saw no point in Daddy's insistence that we stay constantly in his presence so that he could teach us by his example. He loomed over us like a dark behemoth, instilling fear and demanding absolute adherence to all his edicts, however arbitrary they might appear to be. In my first year of law school, I read Oliver Wendell Holmes's dissenting opinion in *Southern Pacific Co. v. Jensen*. "The common law is not a brooding omnipresence in the sky," Justice Holmes wrote, "but the articulate voice of some sovereign or quasi-sovereign that can be identified." I thought at once of Daddy, the brooding omnipresence of my childhood and youth. "He could make me cry just by looking at me," Pigeon often said. But as I grew older, made my own way in the world, and raised a son, I came to appreciate what I had not understood as a child: I had been raised by the greatest man I have ever known.

2

AS GOOD AS US

My grandparents and I were never in any doubt about what I would do after graduating from St. Benedict's. St. Pius X was Savannah's only Catholic high school for blacks, and it was a ten-minute walk from our house. I had dreamed of going there, and I was excited to don the school uniform of gray slacks, white shirt, gold tie, and burgundy blazer. Franciscan nuns also taught at St. Pius X, and not only were they as demanding as the nuns of St. Benedict's, but they assigned far more homework. I often studied at the Carnegie Library until it closed at nine o'clock, then came back home and put in another hour or two at the kitchen table while listening to R & B on WSOK, the local black radio station. When Daddy and I delivered fuel oil late on winter nights, I would ask him to wake me at four or five the next morning so that I could study for a couple of extra hours before breakfast.

I liked St. Pius X, though I didn't have much use for "checking," the all-too-frequent ritual in which students made cruel fun of their classmates' families, clothes, and looks. Most of the insults aimed at me had to do with the darkness of my skin, the flatness of my nose, the kinkiness of my hair, and the way I talked. (My speech was still full of the Geechee dialect I had grown up hearing in Pinpoint and from Daddy and Aunt Tina.) It was only adolescent hazing, but it still hurt. In those days it was an insult to call a dark-skinned Negro black, and more than once when our teacher was out of the room, someone would call me "ABC—America's Blackest Child," an epithet that made many of my classmates roar with laughter. Such racial slurs stung all the more for having come from my own people.

I joined the school paper during my sophomore year, and later I attended a journalism seminar at Savannah State College. I was so impressed by the school that for a time I imagined I might go there and become a newspaperman. But having long been one of St. Benedict's most dependable altar boys, I'd also been thinking vaguely about the possibility of becoming a priest. I waited patiently for a dramatic revelation that would make the right choice clear, but it never came. Then, in 1964, the Diocese of Savannah held a convocation of local altar boys that included an afternoon of activities at Saint John Vianney, the diocesan minor seminary located on Isle of Hope, a pretty island town southeast of Savannah. (*Cape Fear*, *Forrest Gump*, and *Glory* were all filmed there.) Visiting the seminary caused me to start thinking seriously about the religious life, and that spring, a few months shy of my sixteenth birthday, I decided that I wanted to enter Saint John Vianney to prepare for the priesthood.

So far as I know, there were no desegregated schools in Savannah, public or private, in 1964. But change was in the air, as well as in the hearts and minds of the blacks of Savannah, who had started protesting against segregated lunch counters, whites-only parks, and job discrimination. As a result, the public library and some of the public parks were now open to us. The local schools, however, remained segregated—including Saint John Vianney. While the nuns of St. Benedict's and St. Pius X refused to give in to the prevailing stereotypes about blacks, the Catholic Church's official position on segregation was murkier. It maintained separate parishes and parochial schools for blacks, and the local church fathers seemed in no hurry to lift the color bar. No doubt they thought they were being prudent, but before long I would come to see their caution as cowardice.

At the time, though, I had a more immediate concern: I had to tell Daddy what I wanted to do. Much to my surprise, he listened calmly, perhaps because I'd laid the groundwork for our talk by speaking to Aunt Tina, who had undoubtedly broken the news to him. The fact that Saint John Vianney had yet to admit a single black didn't worry Daddy, who was an active member of the local chapter of the NAACP and had routinely put up his property as bond to bail student protesters out of jail. He was more concerned about the expense of sending me to Saint John, which charged about $400 a year, four times as much tuition as St. Pius X. For him that was serious money. Yet he agreed, laying down one condition: "If you go, you have to stay. You can't quit." I promised I wouldn't. He gave me his blessing, then added, "Don't shame me—and don't shame our race." After talking it over with my teachers and our parish priest, I sent in an application. The answer came back at once: Saint John Vianney would admit me, but only if I agreed to repeat the tenth grade, since I hadn't studied Latin at St. Pius X. (Seminarians normally entered Saint John Vianney in the ninth grade and were required to have four years of Latin to graduate.) I was taken aback by the requirement, but didn't let it bother me.

It may have been a blessing in disguise that I didn't know what I was letting myself in for by going to a white school. Blacks in Savannah rarely came into contact with whites, and when we did, the encounters were usually brief and not too unpleasant, since our second-class status was so firmly accepted that no unpleasantness was needed to enforce it. Except for Aunt Tina, most of the black women I knew worked for whites as maids, cooks, or cleaning ladies. They caught the bus early each morning and returned in the afternoon, going to the back to find a seat. We took that slight for granted, just as we accepted the separate water fountains at which we drank in stores, or the fact that we weren't permitted to use certain parks, schools, restaurants, movie theaters, and libraries, or go to Savannah Beach unless we worked there. (Daddy said that white men didn't want black men looking at half-naked white women.) Bit by bit the peculiar institution of slavery had evolved into the peculiar institution of segregation.

I knew all this, but I didn't yet understand it, just as I didn't understand the long, complicated history of race relations in the Deep South. All I knew of the civil-rights movement was what I had read in library books and seen on the evening news. I had no notion of what would happen if I dared to leave the comfort zone of segregation and test the uncharted waters of the larger world.

ON A PLEASANT Sunday afternoon in the fall of 1964, Daddy drove me out to Saint John Vianney. Pine trees lined both sides of the long driveway. To the left were baseball and football fields, to the right the seminary building, in which the dormitory was located. It was only the second time I'd been there, and though it was close to Pinpoint, I might as well have been on another planet. Daddy helped me unload my bags and drove off. I walked into the seminary building, opened the door, and saw a sea of strange white faces. I thought they were all staring at me. I wanted to turn right around and go back home, but it was too late. Later I saw another black student, Richard Chisolm, whose sister had been one of my classmates at St. Benedict's and St. Pius X. Judging from the look on his face, he was in at least as much shock as I was.

I continued to get occasional stares during the first few weeks of school. Sometimes I wondered if certain of my classmates were surprised that I didn't have a tail. But their interest in me, though it made me uncomfortable, didn't seem hostile, and no one treated me badly or showed any signs of outright bigotry—yet. In any case I was too busy learning the ropes of seminary life to be bothered. A bell woke us at six, and a second bell rang at 6:25 to warn us that we had five minutes to get to chapel for morning prayers and Mass. Then came breakfast, after which we made our beds, cleaned up the building, and went to our first class at eight-thirty. The rest of the day was filled to overflowing with structured activities: more classes, manual labor, intramural games, meditation, more prayers, supper, kitchen duty, study hall, evening prayers. Around ten o'clock one of the priests came by the dormitory to give us a blessing and turn out the lights. We didn't get much free time, and what we did get was strictly rationed. Only on Thursday afternoons were we allowed to leave the school grounds (we had classes on Saturday).

The panic I felt on my first day gave way to a constant state of controlled anxiety. My new teachers assigned even more homework than the nuns at St. Pius X, and the classes required more preparation. I wasn't used to that kind of pressure, and I started to worry about flunking out. Such fears were part of the price that my generation of blacks paid for moving out from behind the wall of segregation. We had always believed that we could do as well as whites if we were only given a fair shake—but what if it turned out that we weren't good enough after all? What would that do to our long-deferred dream of equality? What made the difference for me was that I was no longer spending my evenings and weekends delivering fuel oil, or living in a household of which I was the most literate member. Instead I was surrounded by well-educated people, and the mandatory, closely monitored study halls forced me to develop better study habits. Propelled by fear and excitement, I started reading in closer detail and with deeper understanding than ever before.

Grades were handed out every six weeks. To get them we had to go to Father William Coleman, the rector, who gave them out personally. I had never known anyone like Father Coleman, a bright, devout priest from Connecticut whose brisk manner intimidated me. I remember sitting outside his office, wondering whether my path to the priesthood was about to be cut short. As it turned out, I didn't have to worry—my grades were more than good enough—but that didn't make the meeting any easier. Father Coleman told me matter-of-factly that I didn't speak standard English and that I would have to learn how to talk properly if I didn't want to be thought "inferior." He offered to help me improve my speech, suggesting that I tape-record him and some of the other priests. His blunt words hit me like a slap in the face: I thought he was saying that I was inferior because I was black. I was stunned, so much so that I went to the chapel to pray about it. In fact he was treating me the same way he treated many of the other students—years later I found out that he'd said similar things to white students whose accents were about as thick as mine—but his candor hurt me, and it also made me self-conscious about talking out loud in class. Yet it motivated me, too: I vowed that day that no one would ever again say such things to me.

It was natural for me to go to the chapel to pray about my meeting with Father Coleman. The Blessed Sacrament—the real presence of Jesus Christ, body, soul, and divinity, with the physical appearance of bread—was kept there, and seminarians would go to the chapel to pray at all hours. I did so whenever I got upset or confused: being in a holy place always seemed to calm me down and put me at ease. Now that I was free at last to play the team sports of which Daddy had always disapproved, I also found comfort of a different kind on the playing fields of Saint John Vianney. By all accounts I was an outstanding athlete, blessed with speed, quickness, and strength. My sports career wasn't without its rough spots, though. The most painful one was the evening I went out with three other students to play pickup basketball. We decided to shoot free throws to choose teams. I made the first basket, but before I could pick a teammate, the other players moved away from me and started whispering to one another. Another student joined them, and without saying a word to me, they went to the other basket and started playing two on two. I walked back alone to the seminary building, wondering whether they were laughing at me behind my back. I never turned around to see: I couldn't bear to know. Every step was agony. As soon as I got to the building, I went straight to the chapel to pray.

Not long after that, a small group of students was given permission to move the school TV into a classroom to watch the Cleveland Browns play. Those were the days when Jim Brown was on the team. Midway through one of his celebrated runs, one student yelled, "Look at that nigger go!" I felt as if my soul had been pierced. A few of my classmates looked at me shamefacedly, but others snickered. At least two showed every sign of being proud of themselves. I left the room, visited the Blessed Sacrament, and did what I could to compose myself. Worse yet was the time when the same student who'd called Jim Brown a "nigger" passed me a folded note during history class. "I like Martin Luther King . . ." it said on the outside. I unfolded the piece of paper. Inside was a single word: ". . . dead."

Despite these painful incidents, my grades improved steadily, and once I started getting consistent A's in Latin, I knew I was getting somewhere. Toward the end of the first term, I even won a Latin bee. The prize was a statue of St. Jude, the patron saint of hopeless cases. I smiled to myself as I placed the statue on top of my chest of drawers, suspecting that many people must have thought that my efforts to learn Latin would prove hopeless. I went back to the dormitory shortly afterward to find the head broken off. I glued it back in place, but a few days later someone broke it off again. Those who knew the culprit's identity remained smugly silent, but their eyes gave them away. I glued the head back on once more, and this time it stayed fixed. I've taken that statue with me to every job I have held in my adult life, including the Supreme Court.

At the end of my sophomore year, I asked the other seminarians to sign my yearbook. One senior wrote, "Keep on trying, Clarence, one day you will be as good as us." That made no sense to me, so I asked my history teacher what it meant. "You know what it means," he said, looking at me in disbelief. I was still perplexed. Was he talking about the fact that I was a mere underclassman, or about my race? I never knew for sure. I remembered that inscription forty years later when I went to a school reunion. One of the other attendees took me aside and said, "Clarence, you taught me that someone who didn't look like me could be a better student, a better athlete, and a better seminarian than I was. Ever since then I've treated people only as individuals, not as members of groups. Thank you for teaching me that." His words touched me to the heart. If only someone had said that to me in 1965!

Before going out to the farm that summer, I stopped by St. Pius X to tell the nuns how I was doing. I mentioned that I needed to learn how to type, and one of them loaned me a manual typewriter and an instruction book from the school's typing classroom. In addition to learning how to type, I spent what free time I had teaching myself first-year algebra and reading at least five pages of a difficult book every day, underlining all the unfamiliar words and looking them up. My efforts soon bore fruit, but they also gave me the uneasy feeling that I was growing away from my family and friends. Some mocked me for trying to "talk proper" and accused me of thinking that I was better than they were; others asked me to my face whether the fact that I was attending an all-male school made me "funny." Above all, though, it bothered them that I was one of only two black students at Saint John Vianney, and I found their reaction disquieting. If going to a predominantly white school was bad, why were blacks putting their lives on the line to fight for desegregation? My grandparents and most of the older blacks I knew had no problem with what I was doing—they were proud that I was more than holding my own in the white world—but I feared that I could no longer go back to my old life in Savannah, and that I would never be fully at home there or anywhere else. It was around this time that I first read the poems of Robert Frost, one of which captured the way I felt: *Two roads diverged in a wood, and I— / I took the one less traveled by, / And that has made all the difference.* Reading "The Road Not Taken" comforted me as I drifted farther from home. It reflected my feeling that I was the odd man out, anxious and afraid, stripped of alternatives, save for the unthinkable: quitting.

When I returned to Saint John Vianney in the fall, I was stunned to learn that Richard Chisolm had dropped out, leaving me as the school's only black student. On the other hand, more than half of the other students in my class, including the ones who had given me trouble, were gone as well. This made my life easier, as did the requirement that I repeat the tenth grade, which turned out to be one of the best things that could have happened to me. Not only had it given me an additional year to mature, but it also forced me to focus intently on the study of Latin. I couldn't fake my way through class, or pretend I knew more than I really did: I had to know it cold. That kind of discipline was the greatest of the invaluable gifts that Saint John Vianney gave to me. I finally hit my stride, both academically and as a seminarian, and by the time I graduated, my grades were so outstanding that my yearbook photo bore a flattering caption, courtesy of my classmates: ". . . blew that test, only a 98." I treasured that caption—and their friendship—more than any academic prize. Yet I also knew that good grades wouldn't solve all my problems, a lesson hammered home when a priest took some of us to eat at a nearby Big Boy restaurant. The other customers (all white) stared at me with disgust as I walked through the door, and the waitress who served me scowled contemptuously as she shoved my hamburger and fries across the table. Once I would have been content merely to be served. Now I expected to be treated with respect. I didn't get it there, but I did get it now from my teachers and fellow seminarians.

After graduating I spent the summer working as a janitor, groundskeeper, and general handyman at Camp Villa Marie, which was on the same grounds as the seminary. Instead of going out to the farm as usual, I moved into Pigeon's apartment in Savannah, riding the bus to and from Isle of Hope every morning. The $50-a-week salary was my first regular paycheck, and the most money I had ever earned. Though I'd decided to attend Immaculate Conception Seminary, up in the northwest corner of Missouri, I was already having second thoughts about the priesthood, and I briefly fantasized about going to Georgia Tech or West Point. Military discipline didn't intimidate me—life with Daddy had prepared me for anything the Army could dish out—and my success at Saint John Vianney had made me sure of my ability to succeed in any academic setting. I'd always believed that I could do as well as whites, but now I *knew* it: my grades were the proof. Yet hope soon succumbed to reality, since I also knew that it would be all but impossible for a black kid like me to get into either school, and I decided to stick to my religious studies.

IN SEPTEMBER 1967 Daddy drove me to the train station to board the *Nancy Hanks*, the train from Savannah to Atlanta. It was named for a famous racehorse that in turn had been named after Abraham Lincoln's mother, a touch that now strikes me as ironic. Throughout my childhood I'd heard TV ads touting the *Nancy Hanks*, but it had never occurred to me that I'd ride in it someday.

Aunt Tina sent me off with her standard traveling lunch: fried chicken, boiled eggs, several slices of white bread, and pound cake, all packed in a shoe box lined with wax paper. Daddy watched anxiously as I got on the train. He didn't say what he was thinking—that wasn't his way—but I could see his expectations written all over his strong, proud face.

I was one of seven members of the Saint John Vianney class of 1967 who were going to Immaculate Conception, and some of us decided to meet in downtown Atlanta before catching the plane to Kansas City. We had just enough time to look around town, and I especially enjoyed going up and down in the glass elevators of the brand-new Hyatt Regency Hotel. It was the first time I'd ever ridden so high in an elevator, though that was nothing compared to the thrill of my first plane ride. The sudden lift into the air surprised me: I understood the theory of flight, more or less, but that didn't make it any easier for me to grasp the fact that I was actually sitting in a metal tube that had left the ground. As we broke through the clouds, I recalled the words of "High Flight," a once-popular poem that was regularly recited on one of the local TV stations just before it signed off for the night: *Up, up the long, delirious, burning blue / I've topped the wind-swept heights with easy grace, / Where never lark nor even eagle flew; / And while, with silent lifting mind I've trod / The high untrespassed sanctity of space, / Put out my hand, and touched the face of God.* In 1986 President Reagan spoke those lines at the memorial service for the crew of the *Challenger*, and as he did so I thought of that long-ago day in Atlanta, and the thrill I'd felt as I soared through the air for the very first time.

The landing in Kansas City was even more thrilling than the takeoff. We descended along the Missouri River, maneuvered between a building and a tall TV aerial, then plunged to the runway. From the airport we made our way to the downtown bus terminal. There we boarded a bus bound for Maryville, where we were met by a monk who loaded us into a car and drove us to Immaculate Conception. I gazed with wonder at the rolling hills and the vast blue bowl of sky all around me. At last the spires of the basilica at Conception Abbey emerged triumphantly from among the hills.

Unlike Saint John Vianney, which never had more than about sixty students, Immaculate Conception was a good-size campus that was part of a much larger abbey whose monks also worked a farm and a printery. It offered both a four-year undergraduate program and graduate programs in philosophy and theology. Many of the students were adults, and all were required to wear a cassock and Roman collar to morning and evening prayers, as well as to Mass. I bought one of each, and loved the formality (and novelty) of wearing them, the same way I'd loved my school uniforms. For a brief time it seemed that my vocation was becoming more real and substantial, that I was no longer merely playing out a young boy's dreams.

Then I learned that one of my childhood friends had been shot and killed on the streets of Savannah. Tommy, I was told, had quarreled with a young man named Rudolph Alexander, apparently over a girl they both liked. Some of Tommy's friends belonged to a gang called the Tornadoes, and they had attacked Alexander. Not long afterward Alexander lay in wait for Tommy as he walked home from church, and shot him in the chest. He died at the scene. Part of what made Tommy's death so hard for me to take was that Alexander had killed him in front of a store that had long been one of our favorite stopping places. Many of us, including Tommy, used to go there after Mass, since we had to fast for three hours before Communion. Savoring our candy, cookies, or gum, we would make plans to ride our bicycles, play ball, or go see a Sunday matinee together. How could a place of innocent fun have been desecrated in so pointless a way? Was life that cheap? Otis Redding, another Georgian, died that same year, and I still think of Tommy whenever I hear "Sittin' on the Dock of the Bay," Redding's soulful lament of a lonely man who travels from Georgia to San Francisco in search of his destiny, wondering if anything will ever "come my way."

I sought to ease my sorrow by throwing myself into my studies. My classes gave me little trouble, and I got excellent grades from the start, beginning much as I had ended at Saint John Vianney. Before long I was making new friends. One of them, Tom O'Brien, was from Kansas City, and he often took me to visit his family, who welcomed me with open arms. I went home with Tom that Thanksgiving, and we attended a basketball game at a predominantly black high school that Tom thought had the best team in town. Later we met up with several other seminarians to go drinking across the border at a Kansas bar where we bought 3.2 beer, which was then considered nonalcoholic and so could be purchased by underage drinkers. Alcoholic or not, it got us drunk, and the drunker we got, the worse we behaved. I was the only black in the group, and when the management finally decided to throw someone out, they picked me. My classmates all left with me, a gesture of friendship that I never forgot.

The students of Immaculate Conception didn't spend much time getting drunk. We lived by the clock, keeping to a tight schedule like the one to which Saint John Vianney had accustomed me. Our days and nights were filled with study and prayer. Church services at the abbey included Gregorian chants sung in Latin, and I was stirred by the sight of so many monks, priests, nuns, and students deep in songful devotion. I also played basketball, softball, and flag football, doing so well at the school's annual Olympic-type games that some students thought I deserved the superjock trophy traditionally given to the outstanding athlete. No trophy was given out that year. I knew I hadn't come to the seminary to play sports or win trophies, so I swallowed my pride and said nothing, but I couldn't help thinking that I'd been passed over because I was black.

I started spending time with an older student named Dolis Dural, one of the few other blacks at Immaculate Conception. We talked for hours in his room, listening to blues and jazz on his record player. Dolis introduced me to the music of Nina Simone, and he even tried (in vain, alas) to teach me how to play the saxophone. One of our regular topics was the Catholic Church's treatment of blacks. In those days the Church had little to say about racism, which disturbed us greatly. Dolis was the first older black person with whom I discussed such matters. He was well educated and well informed, and the more we talked, the less sure I grew of my vocation. It seemed self-evident to both of us that the treatment of blacks in America cried out for the unequivocal condemnation of a righteous institution that proclaimed the inherent equality of all men. Yet the Church remained silent, and its silence haunted me. I have often thought that my life might well have followed a different route had the Church been as adamant about ending racism then as it is about ending abortion now.

I prayed for guidance in the presence of the Blessed Sacrament, but instead of comfort I found only sorrow and confusion, and by December I'd come to the conclusion that the priesthood was not for me. I spoke to Father Raymond Bane, my parish priest in Savannah, over the Christmas break, and he suggested that I spend one more semester at Immaculate Conception to make sure that I had lost my vocation. He was kind and sympathetic, and while I felt sure that I would never be a priest, I agreed to go back. I said nothing to Daddy—the thought of confronting him was terrifying—but I did talk to Aunt Tina, who told me to do whatever I thought best and promised to help me break the news to Daddy if and when I decided to leave.

I thought of applying to Morehouse College and Fisk University, two of the most prestigious black colleges. I changed my mind when I learned that both schools required applicants to send a photo. It had long been whispered in Savannah that Morehouse and Fisk admitted only light-skinned blacks, and though I don't know whether the rumors had ever been true, they were still widely believed in 1967. I applied to the University of Missouri (and was accepted), but the more I thought about going there, the clearer it became that I wasn't prepared to put myself through the emotional strain of attending yet another predominantly white school. In the end I decided that if I dropped out of the seminary, I would transfer to Savannah State College. As best as I knew, the student body there was all black, and I longed to return to the city I now thought of as my home. Sister Mary Carmine, my chemistry teacher at Saint John Vianney, heard that I was thinking about leaving Immaculate Conception and wrote to suggest that I consider Holy Cross College in Massachusetts. I didn't like that idea any better than going to the University of Missouri, but Sister Mary Carmine, unaffected by my lack of interest, asked Bob DeShay, an old classmate from St. Benedict and St. Pius X who was now a sophomore at Holy Cross, to send me an application. The two of them went to so much trouble that I felt obliged to fill it out, but I put it out of my mind as soon as I sent it off.

I was still torn by indecision when I walked into the dormitory one April afternoon and heard someone shout that Martin Luther King Jr. had been shot. "That's good," another student replied. "I hope the son of a bitch dies." His brutal words finished off my vocation—and my youthful innocence about race. Later that week Dolis suggested that we drive into Kansas City, spend the night with his parents, then participate in a march in Dr. King's honor. We climbed into his Volkswagen and raced to Kansas City to spend the night with his parents. The next day we joined thousands of people in the first of many such demonstrations I would attend over the next few years. As I chanted and sang with the other marchers, I felt a fulfillment that I had never known at Conception Abbey. This was the real world: the seminary was surreal, far removed from the momentous turmoil in which America was now immersed.

I RETURNED TO Savannah in May to face the dreaded task that I'd put off for months. As soon as I got home, I told Daddy that I was dropping out of Immaculate Conception. He listened in a stony silence that was even more terrible than the explosion of wrath I'd feared. When I was finished, he reminded me of my promise not to quit. I told him that he didn't understand what I was going through and that I simply couldn't stick it out. None of it meant a thing to Daddy. He'd never accepted any of my excuses for failure, and he wasn't going to start now. "You've let me down," he said. He didn't have to say so: I knew exactly what our shared dream had meant to him. It justified the sacrifices he'd made and the bigotry he'd endured as he struggled with the daunting obligation of raising Myers and me. Now the dream was a heap of ashes. I had broken my promise, and my failure to live up to my word became a burden on my conscience that I have never escaped.

That night I asked Daddy if I could borrow his car to visit some friends, and he reluctantly gave me the keys. I can't remember where I went, but it wouldn't have mattered. All I cared about was getting out of the house for as long as I could. One of Daddy's strictest rules was that Myers and I be home by midnight, so I was careful to return a few minutes before the hour. The next morning Myers told me over breakfast that Daddy was alone in the garage, weeping over my decision to leave the seminary. I was shocked. Neither one of us had ever seen him shed so much as a single tear. Then he came into the kitchen and said he wanted to see me in the living room. This, too, was unprecedented. Like many southern families, we never used the living room except on the most formal of occasions. A sharp pain gripped my stomach. I knew something bad was about to happen. I had no idea how bad it was going to be.

Daddy sat me down and told me that I'd stayed out past midnight. Even as I denied it, I knew I was wasting my breath: I realized that merely being late wasn't enough to have made him so angry, and I trembled at the thought of what was to come. He said that he'd looked at his clock when he heard me open the front door, and that I'd definitely come home late. His clock, an old wind-up model, was always right, and that was that. His next words seared themselves indelibly into my brain: "No man breaks my door open after midnight." He reminded me of the morning when he had warned Myers and me that the door to his house swung both ways. Now, he said, it was swinging outward. "Because you're acting like a grown man and making decisions like a grown man, you have to live like one. No other man but me will live in this house. I want you to leave."

I was speechless. First the seminary, then the country, now the only real home I'd ever known—all were crashing around me. Where could I go? What would I do? Was it really possible that the man on whom I had always counted was going to turn his back on me? "When do you want me to leave?" I asked weakly. Daddy's answer was like the tolling of a funeral bell: "Today—this day!" I fumbled for something more to say. Would he help me with college? "I'm finished helping you," he said. "You'll have to figure it out yourself. You'll probably end up like your no-good daddy or those other no-good Pinpoint Negroes." The set of his jaw and the steel in his voice left no doubt that his word was final. My life and fate were in my hands.

Reversing the journey I had made almost thirteen years before, I packed the few possessions I'd brought home from the seminary and went to live in Pigeon's apartment. I had no money and no chance of borrowing any, so I immediately started looking for work. I landed a summer job as a proofreader at Union Camp Corporation, a paper company known for the pollution it spewed into the air and water around Savannah. My assignment was to assist the man who proofread the text that was printed on the paper bags manufactured there, and make sure that the materials used to print the bags were in proper condition. Except for one older janitor, I was the only black employee in that part of the plant.

Going to the bathroom at Union Camp was an ordeal. The word "nigger" and the initials "KKK" were scribbled on the walls and carved into the wood of the toilet stalls. I tried not to think about the cold, menacing glares of the men who sat on benches along the bathroom walls, smoking and talking during breaks that seemed far too frequent and too long. None of them ever confronted me directly, though. Instead they ignored me, even when I was talking to them. Some of the other employees weren't openly hostile, but they still treated me differently. Most made it clear that they regarded me as an inferior; others condescended to me, masking their contempt with elaborate displays of kindness, sympathy, or compassion. The contrast reminded me of Daddy's explanation of the difference between rattlesnakes and water moccasins. Both, he said, were deadly, but the rattlesnake was easier to spot. It rattled before it struck, while the water moccasin would strike silently and without warning, making it more dangerous. Like so many of his lessons, I would often have occasion to recall it later in life.

I usually got up around six in the morning, tiptoeing through Pigeon's bedroom into the kitchen to prepare the sack lunch I took with me to the factory. One day I turned on the radio while making my sandwich and heard that Bobby Kennedy had been shot. I fell to my knees and burst into tears. I didn't want to go to work after that, but I knew I had to anyway. No one was going to take care of me, or any other black person in America. At long last I felt the blind, self-destructive rage that haunted so many of the people I knew—even Daddy, who had fought long and hard to keep it in check. Once, I'd asked him why he had decided to start his own business instead of working for someone else. He told me that before daybreak one morning in the late forties, he had been delivering ice when a white man had walked up behind him, startling him. "What're you doing, boy?" the man asked. "None of your business," Daddy replied, clutching the handle of the ice pick he kept in a holster on his belt. "Something boiled up in me," he told me, adding that if the man had made one move toward him, he would have stabbed him to death. After that, Daddy said, he decided he could never work for a white man again.

Now I knew how he felt. Now I understood why he needed a drink when he came home from the humiliating task of getting his business license, when he received a traffic ticket for the fabricated violation of driving with too many clothes on, or when a white woman called him boy in front of Myers and me. Every southern black had known such moments, and felt the rage that threatened to burn through the masks of meekness and submission behind which we hid our true feelings. It was like a beast that lay in wait to devour us. Some fought it with drink, others with prayer. You can hear the struggle in the soulful wails of gospel singers and the passionate moans of blues singers.

I lost my battle with the beast in the summer of 1968. It isn't hard to see why. My family, my faith, my vocation, the heroes who inspired me: all had been taken from me. Once they had helped keep the beast at bay. Now it slipped its leash and began to consume me from within. I began to fear that I would never climb out from under the crushing weight of segregation. No matter how hard I worked or how smart I was, any white person could still say to me, "Keep on trying, Clarence, one day you will be as good as us," knowing that he, not I, would be the judge of that. The more injustice I saw, the angrier I became, and the angrier I became, the more injustice I saw, not only at Union Camp but everywhere I went. I thought a lot about Daddy that summer. He was devoutly religious, honest to a fault, loved his country, and worked harder than anyone else I knew. He did everything right—and where had it gotten him? "I've had to struggle all my life to keep a roof over our heads, clothes on our backs, and food on our table," he often said. In a land of wealth and prosperity, what kind of reward was that? Daddy didn't complain—he never did—but I couldn't accept the way the white man had treated him. Somehow, some way, he and others like him had to be avenged.

Bob DeShay came home from Holy Cross that summer, and we spent long hours talking about the condition of blacks in America. He told me about the theory of Marxism and a new organization called Students for a Democratic Society. I didn't understand everything he was saying, but I got the main point, which was that northern blacks were more radical and confrontational than the ones among whom I had grown up. We fought to cage the beast, while they turned it loose and let it roar. That was the "long, hot summer" of urban riots and nationwide protests, and the more I read about the black power movement, the more I wanted to be a part of it. What was the point of working within the system? Segregation, lynchings, black codes, slavery: the endless litany of injustices raced through my head. Surely the time for politeness and nonviolent protest was over. Look what it had done for Dr. King and Bobby Kennedy—not to mention Daddy, Aunt Tina, and the millions of other compliant, self-deluded blacks who played by the rules. Might it be that those rules were nothing more than a sinister invention devised by the white man to fool blacks into cooperating with the oppressive machinery of American life?

That summer I tore off the beliefs I had learned from Daddy and the nuns, the same way Clark Kent tore off his suit. The fog of confusion lifted. I knew what was wrong, who to blame for it, and what to do about it. I was an angry black man.

3

THE CORRIDOR

When Daddy threw me out of the house, I started thinking seriously about transferring to Holy Cross. I had to. Indecision was a luxury I could no longer afford, and while I hadn't changed my mind about not wanting to attend another white school, the idea of staying in Savannah was even less appealing now that my family life was in shambles. I ranked near the top of my class at Immaculate Conception, so Holy Cross had quickly accepted my application. The only problem was money, but the director of financial aid assured me that something could be worked out. At first Daddy refused to help me fill out the forms, but he finally gave in and let me talk to his accountant. I was astonished to discover how little he and Aunt Tina made from their fuel-oil business. In the best years, their gross income never rose above $7,000 a year, and often just a fraction of that. How could they have worked so hard for so little, and without a word of complaint? At one time I would have admired their stoicism, but now I took it as a sign of weakness, a willful refusal to face the realities of racism in America.

Come September I was ready again to head north. Aunt Tina gave me another shoe box full of food and told me to wear clean underwear in case I was hit by a car and never to eat anyone's chitterlings but hers, since nobody else cleaned them properly. I knew how sorrowful my break with Daddy had made her; she knew that there was nothing she could do about it but watch and pray. As for Daddy, he had plenty to say to me, none of it good. I had broken not just with the Catholic Church but also with the principles he had worked so hard to teach me, and he couldn't see why I had turned my back on them. I blamed his failure to share my rage on his lack of formal education, aggravated by years of brainwashing from the white man. He warned me once again as he drove me to the train station that I was well on the way to becoming just like my father and the other "no-good Negroes" in Pinpoint. I swore to myself that I would prove him wrong. As the train pulled out of the station and Savannah vanished in the distance, so did the nice black boy from East Thirty-second Street who had wanted nothing more than to be a priest. By then Holy Cross was far more than just a school to me. It had become the embodiment of all my hopes for the future, my last chance to do more than merely eke out a bare living in the segregated South, the way Daddy had. It would be my escape, my emancipation.

Holy Cross was still an all-male school in 1968. It was also nearly all white: I was one of 6 black students in a class of about 550. Life there was less tightly regimented than at Immaculate Conception. Students were free to leave campus whenever they liked, without permission, and attendance at religious services was not compulsory. At the seminary I had been expected to wear a coat and tie to class, but Holy Cross imposed no such obligation, and many of my classmates wore sandals and cut-off jeans wherever they went. That suited me, since I now believed that the whole of American culture was irretrievably tainted by racism, and I showed my approval of the coming revolution by donning Army surplus clothes and footwear, just as I showed my solidarity with my new black friends by adopting their handshakes and jargon and becoming the corresponding secretary of the newly organized Black Students' Union. Perhaps in part because I had spent the past few years in a mostly white environment, I now found it comforting to be among students who looked like me and felt as I did. Our race, we thought, separated us from whites in ways that only we could appreciate. It gave us a swagger, a sense of moral superiority. We were the aggrieved and the righteous.

In many ways, though, I was still the same hardworking young Savannah boy who had come to Saint John Vianney in the hope of bettering himself. For all my growing disaffection, I continued to earn very good grades, and I was so determined to get all I could out of Holy Cross that I actually decided to major in English literature, since I had yet to fully master the fine points of standard English. Father Coleman's warning that failure to do so would cause me to be seen as inferior had stuck with me. For the same reason, my revolutionary fervor stopped well short of experimenting with the mind-altering drugs that were readily available at Holy Cross. I thought it safer to drown my sorrows in beer and cheap wine. I got along just fine with the white people I claimed to hate, and I grew close to my white roommate, a pleasant, conscientious young man named John Siraco who became one of my best friends.

Yet my new beliefs had changed me all the same, and one of the biggest changes was that I parted ways with the Church. During my second week on campus, I went to Mass for the first and last time at Holy Cross. I don't know why I bothered—probably habit, or guilt—but whatever the reason, I got up and walked out midway through the homily. It was all about Church dogma, not the social problems with which I was obsessed, and seemed to me hopelessly irrelevant. So did the athletic events in which I now took part only sporadically. Once I'd found them important and fulfilling, but now I knew they were nothing more than another way for "the man" to oppress black people. Racism had become the answer to all my questions, the trump card that won every argument. I was furious with the Church, with Daddy, and with the condition of blacks in America, and now that I was surrounded by other students who shared my fury, I began to think of myself as a man without a country. At the same time, I was haunted by Daddy's prediction that I would end up just like my father, and by the dire warnings of the friends and family members who told me over and over that "the man ain't goin' let you do nothin'. Why you even tryin'?" Each day I found myself fighting off the urge to give up and return to a simpler, easier existence—but then I would remember trudging up and down the fields of the farm in midsummer, overwhelmed by heat and tormented by gnats and mosquitoes, and decide to stick it out.

I spent the first part of my summer vacation working two jobs, one at an electroplating company in Worcester. It was an unhealthy place—the air was full of evil-smelling fumes—and near the end of the summer recess I became too sick to continue working. I took a bus to Savannah to visit my family and recover my health. That was when I saw how much I'd changed. Daddy and I never seemed to stop fighting. I complained bitterly about the oppression of blacks and told him that a revolution was coming. He assured me in return that America was the best country in the world, but I stood my ground and argued right back at him, and my newfound insolence made him furious. "I don't know why I worked so hard to help you, boy," he said more than once. "I never thought you'd go to some damn school way up north and have all this foolishness put in your head." By the end of my stay, he would simply leave the room whenever I started "talking crazy," which made me madder than ever.

Ever since I was a boy, I had looked up to Daddy as the patriarch of our family and the ultimate source of authority in my life. He knew how to build a house, deliver fuel oil, and put food on our table and clothes on our backs. But he'd never heard of Hegel, Kierkegaard, or Marx, so I wrote him off as an ignorant illiterate incapable of understanding or facing the facts about racism. How could a black man from the Deep South who had survived the worst kind of bigotry possibly refuse to admit that America was tainted and corrupt and had to be rebuilt from the ground up? Was he a coward, or just a fool? Aunt Tina looked on in anguish as we fought, nervously winding and unwinding her handkerchief. She begged me to stop fighting with Daddy, assuring me that he had done his best to raise Myers and me right and that he loved us deeply. I told her that I could see no evidence of his love and no reason for me to stop arguing. Neither could Daddy. He never gave an inch.

I FELT SURER of myself when I returned to Holy Cross that fall. Though my first year there had been tougher than I'd expected, I was a dean's list student, and I proudly informed one of my black classmates that I wanted to go to Harvard Law School. He laughed, but I set my jaw and told him I was serious. I wasn't, not really. Harvard itself meant nothing to me compared to the prospect of helping to right the wrongs of segregation. The thought of going there wasn't much more than an adolescent fantasy, but it had the advantage of being both tangible and ambitious. It wasn't easy being black at Holy Cross, and without a clear-cut goal to strive for, no matter how unrealistic, I might well have floundered and gone under, as so many of my new friends were to do.

To my knowledge there was no significant difference in the academic records of the black and white students at Holy Cross, and many of the blacks who went there did superlatively well. My friend Gil Hardy, for instance, was a seventeen-year-old freshman from Philadelphia whose slangy talk and self-deprecating, down-to-earth demeanor fooled some of his classmates into underestimating him—though not for long. Even as a freshman, Gil took mostly upper-level classes, including Greek and Latin, and his first-semester grade-point average of 3.9 (he got one B+) won him the nickname Three-Nine. Nor was the administration unaware of the difficulties that we faced as the school's first group of black students. Father John E. Brooks, the vice president for academic affairs, was especially sensitive to our situation, and though he refused to water down the school's stringent academic requirements, he did everything he could to help us meet them.

But for every Gil Hardy, there was another talented black who was losing his way at Holy Cross, and I soon saw that merely being smart was no guarantee of success. Some black students gave up and stopped going to class, while others started using drugs or dabbling in cultlike Eastern religions. Their problem was that they lacked the social experience that would have made it easier for them to leave the comfort zone of segregation and move into the white world. Many of them, I suspected, might have done better had they gone to schools closer to home or to predominantly black colleges, which would have allowed them to grapple with the ordinary challenges of young adulthood without having to simultaneously face the additional challenge of learning how to live among whites. Yet Holy Cross, like other colleges across the country, continued to admit them in fast-growing numbers. When I arrived, there was only one black senior and two juniors; I was one of six blacks in my class, and seventeen were admitted in the freshman class. Too many of the latter group did poorly, as did subsequent classes, and some failed outright. I couldn't see the point of putting them through an experience for which they were unprepared. Why, I asked, were these gifted young people being sacrificed on the altar of an abstract theory of social justice—and who profited from their failure?

That was my first brush with racial heterodoxy. The next one came when the members of the Black Students' Union voted to set up a separate black living area known as the "black corridor." Supporters of the plan claimed that because there were so few blacks at

Holy Cross, it was important that they live together so as not to feel isolated. I didn't see it that way. Did we really want to do to ourselves what whites had been doing to us? Besides, I liked my white roommate and didn't want to stop living with him. But the other members of the BSU voted for the corridor, and in the fall of 1969 the administration allowed black upperclassmen to live together on the fourth floor of one of the dormitories. For the sake of "solidarity," I chose to live there instead of going my own way—so long as I could keep my roommate. John, bless him, went along with my scheme, and our friendship remained intact.

Not all of Holy Cross's black students moved onto the corridor. Some continued to live off campus, while others objected to the corridor on principle and were ostracized for refusing to live there. I secretly admired their tenacity. I had already started to notice that many of my fellow blacks found it hard to relate to white students other than confrontationally, and I suspected that the existence of the corridor would make it harder for them to adjust to life at Holy Cross. Even though I was not intimidated by whites, I still felt the tension that arose from my unfamiliarity with white customs, and it may be that the corridor helped me and other black students to deal with this chronic and predictable problem. Still, I knew we couldn't have it both ways, at least not for very long. Sooner or later we would all have to learn how to live among whites, and I saw no reason to put it off any longer than was absolutely necessary.

That wasn't the only thing I disliked about the the corridor. I was also troubled by the alacrity with which Holy Cross had yielded to our demands. Some blacks on campus already thought that the mere existence of racial oppression entitled them to a free pass through college, and the administration's apparent willingness to accommodate us now led these black students to assume that they would always be able to get whatever they wanted. But I foresaw a time when it would no longer be fashionable to give blacks a helping hand, especially after the generation of whites who remembered segregation was gone, and it seemed just as clear to me that Hispanics and women would soon start making similar claims, thus putting them in competition with blacks.

Preferential policies intended to help blacks adjust to life after segregation were very much on my mind in those days, and now I began to think them through in a more systematic way. Talented blacks stuck on the bottom rung of the socioeconomic ladder clearly deserved such help, but the ones who most often took advantage of it were considerably higher up on the ladder. Most of the middle-class blacks with whom I discussed these policies argued that all blacks were equally disavantaged by virtue of their race alone. I thought that was nonsense. Not only were some blacks more economically successful than others, but many light-skinned blacks believed themselves to be superior to their darker brethren, an attitude that struck me as not much different from white racism. Even now blacks don't like to talk about that kind of prejudice, but it had been a very real part of my life in Savannah, which was for all intents and purposes segregated not merely by race but also by class and color. I thought that preferential policies should be reserved for the poorer blacks whose plight was used to justify them, not the comfortable middle-class blacks who were better prepared to take advantage of them—and I also thought the same policies should be applied to similarly disadvantaged whites.

On the other hand, I didn't think it was a good idea to make poor blacks, or anyone else, more dependent on government. That would amount to a new kind of enslavement, one which ultimately relied on the generosity—and the ever-changing self-interests—of politicians and activists. It seemed to me that the dependency it fostered might ultimately prove as diabolical as segregation, permanently condemning poor people to the lowest rungs of the socioeconomic ladder by cannibalizing the values without which they had no long-term hope of improving their lot. At the time, these ideas seemed to me a logical extension of my distrust of "the man," though in fact they were rooted in the lessons Daddy had taught me. I didn't yet know how heterodox they were, much less that they were about to lead me away from the radical politics in which I thought I believed.

AL COLEMAN, A St. Pius X alumnus who had become one of my closest friends at Holy Cross, was as angry as I was, and even more impetuous. Though he was extremely bright, he found it hard to adjust to campus life, and cut classes regularly. Al was already close to flunking out of school when he and several other black students decided to take part in a protest against General Electric's on-campus recruitment program. (GE was a defense contractor, making it a favorite target of the antiwar movement, and reportedly also did business in South Africa.) The protest was in violation of school rules, and the dean of students had the protesters photographed so that they could later be identified and disciplined.

The BSU took it for granted that the black demonstrators would be singled out for punishment, and when five of them, including Al, were duly suspended, we swung into action. Holy Cross, it seemed, was a microcosm of America, with separate rules for blacks and whites. This was intolerable to us, and we were determined to make our anger known—though not in a destructive way. I was one of the BSU members who supported the idea of simply leaving a place where we no longer felt welcome, and that was what we ended up doing. We put on our best clothes, packed our bags, gathered in the ballroom of Hogan Campus Center, announced that we were quitting school in protest, and marched out.

As I got ready to head home to Savannah, I started thinking about what I would tell my grandparents. Suddenly it hit me that I was in deeper trouble than I'd thought. Daddy had always said that there was no good reason for blacks to leave the South, where they were part of the native culture. Going to Immaculate Conception to become a priest was one thing—that made sense to him—but when I abandoned that goal, he saw no point in my moving on to yet another northern school. He expected me to come home in defeat, just like the other members of his family who had emigrated to the North, then slinked back to Georgia "with their tails between their legs." I shuddered at the thought of telling him that he was right. Facing him would be like staring into a mirror: I wouldn't be able to turn away from the reality of what I'd done to myself. But how could I escape it? What alternative did I have?

Within a few hours my problem was solved. Art Martin and Ted Wells, the leaders of the BSU, persuaded the administration to let us return to campus, and the suspensions of the five protesters were subsequently lifted. Al promised that he would stop cutting classes, but he never did, and before long he was kicked out of school, this time for good. I never saw him again, though I think of him often. He was among the first of many young black men who came to Holy Cross full of hope and left it in despair, victims of a well-intentioned theory that failed to take into account the realities of black life. Their fate enraged me—but I had no desire to join their ranks. What good would it do? To give up now would be unforgivable, knowing as I did what Daddy and Aunt Tina had given up on my behalf. I owed it to them to walk through the doors of opportunity that Holy Cross had flung open before me. I returned to my dorm room knowing that I would never again leave Holy Cross until I graduated. I will forever be indebted to Art, Ted, and the school adminstration for giving me a second chance.

The beast of rage kept on gnawing at my soul, but the more I saw of radicalism, the more I doubted that it had any answers to offer me, especially after a Black Panther from Boston came to Holy

Cross to meet with a group of black students. He treated us like children, telling us we were wasting our time at school and that the only thing "the man" understood was a gun. Then he pulled one out of his pocket. I wasn't impressed by the gesture, or by his flamboyant rhetoric. As much as I hated the injustices perpetrated against blacks in America, I couldn't bring myself to hate my own country, then or later.

I hadn't quite turned my back on left-wing politics, though. In the spring of 1970, I was one of several BSU members who went to Boston to take part in an antiwar rally. Holy Cross had previously loaned the BSU a station wagon, since few black students owned cars and thus were unable to travel to other schools to participate in academic or social events, but they never imagined that we would use it to attend an out-of-town demonstration. Nor did any of us suppose that there would be more to this one than the usual inflammatory speeches—but we were wrong. Once the organizers of the rally had gotten the crowd sufficiently worked up, they urged us to march to Harvard Square to protest the treatment of America's domestic political prisoners. Off we went, chanting "Ho, Ho, Ho Chi Minh" and demanding freedom for Angela Davis, Erica Huggins, and anyone else we could think of. When we came to a liquor store, the owner, fearing that we would smash it up, gave some of us the wine we wanted for free. From there we drank our way to Harvard Square, where our disorderly parade deteriorated into a full-scale riot. The police fired rounds of tear gas into the crowd, but that didn't deter us, and we kept on rioting well into the night. Eventually the disturbance fizzled out, and my friends and I went off to look for our station wagon. Somewhere along the way, we ran into a group of policemen wearing riot gear and wielding billy clubs. "This must be the nigger contingent from Roxbury," one of them said loudly. We ignored him and kept on walking until we found the car.

I got back to campus at four in the morning, horrified by what I'd just done. I'd put my academic career on the line by participating in the BSU walkout a few months earlier, but at least I'd known what I was doing and why I was doing it. This was different: I had let myself be swept up by an angry mob for no good reason other than that I, too, was angry. On my way to breakfast, I stopped in front of the chapel and prayed for the first time in nearly two years. I promised Almighty God that if He would purge my heart of anger, I would never hate again.

I began to suspect that Daddy had been right all along: the only hope I had of changing the world was to change myself first. I thought of the many times that he and I had delivered fresh-picked farm produce to one of our elderly relatives. On such occasions he never failed to remind me that if we hadn't worked so hard to grow it, we wouldn't be able to give it to those who needed help. For the past few years, I'd been sneering at the simplemindedness of his philosophy of self-reliance, but now it was making sense to me again. If I was truly serious about helping other people, I'd have to start by helping myself, and the first thing I had to do was chain the beast of rage and resentment that threatened to wreck my academic career and my life. Of course I had every reason to be outraged by the experience of blacks in America, but I had no right to confuse their collective sufferings with my own personal experiences. Daddy had never been able to understand how I, a college student, could consider myself "oppressed." He didn't think of himself that way and didn't see why I should. My job, he insisted time and again, was to "play the hand you're dealt," the way he'd done his whole life. Besides, I had a better hand than he'd ever held—and we both knew it. My life was full of opportunities of which he had never dared to dream. All I had to do was reach out and take them. What right, then, did I have to whine about "the man"? Once, I would have argued him into the ground, but now I had no answer.

The riot in Harvard Square was the last such event in which I took part. After that I stuck to my studies. I'd been drunk on revolutionary rhetoric, but now I knew it was nothing more than talk. Doing well in school seemed small and insignificant by comparison, yet it had to be done. It was my only hope.

BY THE END of my junior year, I felt more comfortable at Holy Cross than ever before. My renewed efforts to excel academically were paying off: I was accepted into the honors program, and I became a member of Alpha Sigma Nu, the Jesuit equivalent of Phi Beta Kappa. I also became a resident assistant on the black corridor, which entitled me to a room of my own and replaced my previous work-study job in the dining hall kitchen. I couldn't have asked for a better roommate than John Siraco, but I thought it was time to try living alone, not only because I'd never done it before but because it would allow me to study even harder. Experience had taught me that I did my best work in the early mornings, and once I started living on my own I was able, as Daddy had, to get up between three and four without having to worry about disturbing John. As I settled into my new schedule, I grew more confident of my ability to learn whatever my teachers threw at me. Even though I'd always gotten excellent grades, I was still nagged by the fear that I wouldn't be able to learn everything I needed to know in order to do the things I wanted to do. Now I became genuinely excited about my classes, and felt the thrill of true intellectual growth. Though my literature classes continued to test my abilities, I even began to think that the day might someday come when I would learn to love the classics. Perhaps there really were better reasons to learn than mere necessity.

The more I read, the less inclined I was to conform to the cultural standards that blacks imposed on themselves and on one another. Merely because I was black, it seemed, I was supposed to listen to Hugh Masekela instead of Carole King, just as I was expected to be a radical, not a conservative. I no longer cared to play that game. Some of my friends accused me of being contrary for its own sake, but I knew there was more to my growing skepticism than mere stubbornness. The black people I knew came from different places and backgrounds—social, economic, even ethnic—yet the color of our skin was somehow supposed to make us identical in spite of our differences. I didn't buy it. Of course we had all experienced racism in one way or another, but did that mean we had to think alike?

It was around this time that I read Ayn Rand's *Atlas Shrugged* and *The Fountainhead*. Rand preached a philosophy of radical individualism that she called Objectivism. While I didn't fully accept its tenets, her vision of the world made more sense to me than that of my left-wing friends. "Do your own thing" was their motto, but now I saw that the individualism implicit in that phrase was both superficial and strictly limited. They thought, for instance, that it was going too far for a black man to do *his* thing by breaking with radical politics, which was what I now longed to do. I never went along with the militant separatism of the Black Muslims, but I admired their determination to "do for self, brother," as well as their discipline and dignity. That was Daddy's way. He knew that to be truly free and participate fully in American life, poor blacks had to have the tools to do for themselves. This was the direction in which my political thinking was moving as my time at Holy Cross drew to an end. It went without saying that I was an individual: we are all individuals. The question was how much courage I could muster up to express my individuality. What I wanted was for everyone—the government, the racists, the activists, the students, even Daddy—to leave me alone so that I could finally start thinking for myself.

In the last semester of my senior year, I took an independent-study course on black novelists. Richard Wright's *Native Son* had already made a deep impression when I'd read it in high school, but now it meant even more to me. As I reentered the nightmare world of Bigger Thomas, the innocent young black man who finds himself caught up in a chain of circumstances that spins out of control and causes him to commit an act of violence that leads inexorably to his own death, I envisioned myself slipping into a similar vortex of self-destructive behavior. Something like that, I realized, could have happened to me in Harvard Square. I resolved to lead my life in such a way as to steer clear of such potentially deadly situations.

I'd read Ralph Ellison's *Invisible Man* in high school, too, but was too young to understand it. Now it made perfect sense to me. Every character in the novel, I saw, was either acting out a stereotype or responding to one:

> I was never more hated than when I tried to be honest. Or when, even as just now, I've tried to articulate exactly what I felt to be the truth. . . . I was pulled this way and that for longer than I can remember. And my problem was that I always tried to go in everyone's way but my own. I have also been called one thing and then another while no one really wished to hear what I called myself. So after years of trying to adopt the opinions of others I finally rebelled. I am an *invisible* man.

There it was. How could a black man be truly free if he felt obliged to act in a certain way—and how was that any different from being forced to live under segregation? How could blacks hope to solve their problems if they weren't willing to tell the truth about what they thought, no matter how unpopular it might be? I already knew that the rage with which we lived made it hard for us to think straight. Now I understood for the first time that we were *expected* to be full of rage. It was our role—but I didn't want to play it anymore. I'd already been doing it for too long, and it hadn't improved my life. I had better things to do than be angry.

One of the things I most wanted to do was become a lawyer, and in December I received a letter of acceptance from Harvard Law School. I smiled wryly to myself as I remembered the laughter that had greeted my announcement two years earlier that I wanted to go there. From that moment on, Harvard Law had seemed like the fake rabbit at a dog race that wasn't meant to be caught, only chased—but I'd caught it. As soon as I finished reading the letter, I went straight to my room and called my grandparents. I'd never bothered to tell them of any of my previous academic achievements, fearing that they wouldn't understand how much they meant to me, while they in turn had never opened Holy Cross's reports on my academic performance. Now I felt the urgent need to let them know that their sacrifices had not been in vain. Aunt Tina answered the phone, and I stammered out my good news. "That's nice, son, if that's what you want to do," she said gently. "But when you going to stop going to school?" Her innocent words knocked the wind out of me. I saw that she hadn't the slightest idea of what Harvard was, or what it meant for me to be admitted to its law school. To her it was just one more of the places that had filled me with wild talk of revolution.

I paid a visit to Harvard some time later, and didn't like what I saw there. The law school seemed too big and too conservative—and, perhaps, too intimidating. After I got back to Holy Cross, I called the admission's office to decline the offer. The woman to whom I spoke couldn't believe her ears, but I assured her that my decision was final. By then I was more interested in Yale Law School, which was smaller and (so far as I could tell) more liberal. I'd also been accepted by the University of Pennsylvania Law School, but I thought Yale might be a better place for me to grow intellectually. At that point I knew nothing of the big-city law firms that hired Ivy League law school graduates by the truckload. All that mattered to me was to learn as much as I could, get my law degree, and go home to Savannah to become a good lawyer and a community leader. I longed to put the North behind me once and for all.

In April I received a thin envelope from Yale Law School. I knew it had to be a rejection letter, since the acceptance letters I'd received from Harvard and Penn had been bulging with paperwork I needed to complete. As I tore it open, I steeled myself for bad news. I was wrong: Yale wanted me. My heart raced and my spirits lifted. This time, though, I didn't bother to call home. Now that I had been accepted into the school of my choice, all my thoughts were on graduating from Holy Cross—and on my impending marriage.

4

NO ROOM AT THE TOP

Early in my second semester at Holy Cross, I met Kathy Ambush, a freshman at Anna Maria College, a nearby Catholic school for women. We started dating not long afterward, and since she shared my political views, we also went to demonstrations together and helped to launch a free breakfast program, patterned after a similar program run by the Black Panthers, for local inner-city children. Kathy's mother worked in a doctor's office, while her father was a technician in a local dental lab. At first they had their doubts about their daughter's new boyfriend, and I can see why. Back then I usually dressed in Army fatigues and a big hat called an "apple," or a "brim," that I insisted on wearing inside the house. I also drank a lot. In those days I favored Ripple, an inexpensive fortified wine that was popular with blacks and college students, but I put away plenty of brandy, beer, and malt liquor as well. I could tell that Mrs. Ambush was forcing herself to tolerate me, and I know that I didn't look like a promising match for her daughter.

In time, though, the Ambushes accepted me as a member of the family, and I came in turn to love, respect, and admire them. They gave me the guidance and support that I thought I could no longer count on receiving from Daddy or Aunt Tina. Years of working alongside Daddy had made me lose my appreciation of the value of leisure, but the Ambushes brought it back into my life during the countless weekends I spent at their home in Worcester. Mr. Ambush and I liked to chat while working in the yard or tinkering with his car. He set up a place at his basement workbench so that I could study. Several of my teeth had been pulled when I was a boy, and as soon as he noticed that the others were starting to shift, he made a bridge to keep them in place. When I refused to purchase a class ring and a yearbook, he insisted that I buy them and offered to give me the money to do so, knowing I would someday be glad I had them. (He was right.)

Kathy and I got engaged midway through my junior year. We decided to get married the day after I graduated, which would make it possible for my family to attend both ceremonies in a single visit. I knew the trip would be expensive and that Daddy in particular would be reluctant to travel so far from home, but I felt sure that if the two biggest events in my life were to take place within a day of each other, they'd do their best to be there. I was to graduate on June 4, 1971, three weeks before I turned twenty-three and two days before Kathy's twenty-first birthday, so we set the date for our wedding accordingly and told our families. Aunt Tina and my mother agreed to come, but Daddy said no, claiming that he had to stay on the farm and take care of the animals. I told him that it was important to me that he be there and even offered to pay for his ticket, but he was adamant. Perhaps our arguments had caused him to write me off as one of those "educated fools with no damn sense" for whom he had nothing but contempt. Whatever the reason, his refusal was a cruel and hurtful blow, and it deepened the chasm that now separated us.

My mother helped me clean out my dorm room the day before graduation. I took my trunk and other possessions to the barn behind the Ambushes' house to store them until Kathy and I returned from our honeymoon. I'd toyed with the idea of skipping the graduation ceremony altogether, but the Ambushes wouldn't hear of it. (I also considered wearing a dashiki to the ceremony, but I thought better of that one all by myself.) By then I was starting to have serious doubts about getting married, and after the rehearsal I stayed up all night with a friend and tried to drink them away. It wasn't that I had any doubts about Kathy. She was the best person in my life, and her parents were as kind and good as she was. *I* was the problem. One of my relatives back in Savannah had warned me that I still had some growing up to do before I'd be ready to get married. Now I wondered whether I should have taken his advice. At the very least, I needed more time to take in what was happening to me and think it through—but my time was up.

I'd long since fallen out of the habit of good grooming, so I had to spend quite a bit of time the next morning trying to untangle my uncut, unruly hair, finally combing it into an Afro that I parted down the middle for no particular reason. Then I put on my rented tuxedo and went to the church, which was full of familiar faces. It was a beautiful day, mild and cloudless, and Kathy was radiant, but I was still full of doubts, and a bolt of sharp, sickening pain shot through my body as we said our vows. It didn't help that I was hungover, but there was more to it than that: I really didn't know whether I was doing the right thing. For one crazy moment, I thought of stopping the ceremony, but I knew it was too late to turn back, and I wasn't even sure I wanted to change my mind. Being with Kathy had already brought me more joy than I'd ever known in my adult life. How could I walk away from her?

We drove up to Montreal the next day for a brief honeymoon, then moved into an efficiency apartment in the inner city of New Haven, where I spent the rest of the summer working for New Haven Legal Assistance, a job I got through Judge Angelo Santaniello, a Holy Cross alumnus. The point of becoming a lawyer was to help my people, and Kathy was as idealistic as I was. We both felt that it was time to climb down from the ivory tower of college life, roll up our sleeves, and start doing some good. Kathy had dropped out of Anna Maria College a year and a half earlier, so she now found a job as a bank teller, while I went looking for an affordable used car, settling for a 1965 Volvo that cost $650. It had been owned by a young woman who had just graduated from Yale and was moving to Hawaii. I wondered how she could have afforded a new car that we could barely afford to buy used. I worked on that Volvo almost every Saturday morning, setting the timing and changing the plugs, points, and oil, and though I was strictly a shade-tree mechanic, it served us reasonably well for the next seven years.

I learned a lot at New Haven Legal Assistance, not all of which had to do with the law. One day I saw another summer employee, a Yale undergraduate, reading a Kurt Vonnegut paperback as we waited for a meeting to start. I'd heard Vonnegut's name but had never read any of his books, and the young man told me that he was a "black humorist." "I didn't know he was black!" I replied. My ignorance was bad enough, but the smugness with which he then explained the concept of black humor to me was even worse. It was a shaming reminder that while I'd graduated from Holy Cross with honors, I still had plenty to learn—and not just from books. Like so many of my black friends, I'd always put great stock in the ability of grassroots community groups to solve the problems of poverty, but the bickering and incompetent leadership of the local groups with which I worked that summer opened my eyes. I grew more pessimistic, and began to wonder why I was even bothering to go to law school. Would I ever make any difference? I spent hours sitting by myself in our one-room apartment, guzzling blackberry brandy and listening to Marvin Gaye's *What's Going On* on a record player I'd bought from a Holy Cross student. I brooded over the futility of life as I listened to that half-despairing, half-hopeful album, in which Gaye asks whether anyone cares enough to save "a world in despair."

The old fear of failure swept over me again when classes started in September. As hard as English literature had been, learning the law was even harder. Cases and terms of which I knew nothing swirled about me in an incomprehensible miasma. No less puzzling was the way in which some of my new classmates jumped self-confidently into the fray, talking back to the professors as if the tangled complexities of legal doctrine were second nature to them. Where had they learned so much? Would I ever catch up? I set aside fifty hours a week to study, plus another fifteen hours to work at New Haven Legal Assistance, which left me with next to no time for extracurricular activities. Panic and dread threatened to overwhelm me, but I put my shoulder to the wheel and got through the first semester in one piece, making several good friends along the way. Getting to know students like Harry Singleton, a Johns Hopkins graduate who came from a family not unlike my own, and Rufus Cormier, a second-year student who'd been a football star at Southern Methodist University but was no less adept in the classroom, helped to ease my anxieties.

It helped, too, that I was working for New Haven Legal Assistance, which made me feel grounded in my community, something I needed badly in those first nerve-racking months at Yale. I'd already noticed the tension between lawyers who wanted to reform the legal system from the top down and those who concentrated on practical problems, and I preferred the latter. The office where I worked was located in a racially and ethnically mixed neighborhood, and while most of our clients were black, we also helped whites and Hispanics. Pearlie Carter, one of the investigators, became a good friend (and, later, the godmother of my son). Pearlie lived in the housing projects just off Dixwell Avenue, close to my apartment, and she sometimes gave me a ride home after work. She was a proud participant in the federal Work Incentive Program, whose purpose was to move adults from public assistance to self-sufficiency, and she often wore a WIN button (the program's commonly used acronym) to the office. I liked her blunt honesty, but I suspect she thought I was more than a little bit naive, since she took it upon herself to school me about the realities of life in the projects. She knew that our clients didn't always tell the truth, especially when they wanted something from us, and thought I was far too accepting of the weaknesses of the welfare system. Unlike me, she was a harsh critic of its overpermissive use and the destructive effects it had on the people it was supposed to help.

Pearlie and I laughed a lot together. Much of her humor was black in both senses of the word. She told me, for instance, about a woman who came into the office and, while filling out the necessary papers, mentioned that she had a daughter named "Fa-*mah*-lee." Pearlie asked the woman how she spelled it. "F-E-M-A-L-E," the woman replied. Doing her best to keep a straight face, Pearlie asked how she'd come up with the name. "I didn't," she said. "The doctor gave it to her." We laughed, but our laughter was tinged with sadness. We knew all too well what the woman's answer said about the world from which she—and we—had come.

Not long after I finished my first year of law school, Kathy and I moved into a one-bedroom apartment in Yale's married students' housing complex. One day my wallet fell out of a hole in my pocket while I was walking to school. I didn't have much money in it—I didn't have much money, period—but it did contain my driver's license, draft card, and school ID. The registrar's office told me that one of my classmates, a young white man named John Bolton, had found the wallet and turned it in. When I went to thank him for his kindness, I learned to my surprise that he was my upstairs neighbor. That was the beginning of a priceless friendship. John was known as a conservative while I still thought of myself as being far left of center (when I wasn't just being cynical). That didn't bother either of us. I'd already noticed that too many white students at Yale seemed not to want to talk to me about anything but black issues or sports. Not John. He was no panderer: he spoke to me about whatever was on his mind, and he was ready to argue at the drop of a hat.

One of our most fateful discussions began when I mentioned to John that I supported mandatory motorcycle helmet laws and the then-new automobile seat belt requirement. Since society bears the cost of care for people who are injured, I argued, the government had the right, if not the obligation, to take steps to reduce the risk of injury. Naturally John thought otherwise, and we argued back and forth. Then he looked me in the eye and said, "Clarence, as a member of a group that has been treated shabbily by the majority in this country, why would you want to give the government more power over your personal life?" That stopped me cold. I thought of what Daddy had said when I asked him why he'd never gone on public assistance. "Because it takes away your manhood," he said. "You do that and they can ask you questions about your life that are none of their business. They can come into your house when they want to, and they can tell you who else can come and go in your house." Daddy and John, I saw, were making the same point: real freedom meant independence from government intrusion, which in turn meant that you had to take responsibility for your own decisions. When the government assumes that responsibility, it takes away your freedom—and wasn't freedom the very thing for which blacks in America were fighting?

John's question reverberated in my mind for a long time to come, though it didn't stop me from voting for George McGovern shortly afterward. Though he was still a bit too conservative for my taste, he would have to do. Certainly no self-respecting black could ever vote for a Republican—or so I thought.

* * *

ON THE SURFACE Yale Law School was everything I'd hoped it would be. The students were smart, the environment relaxed but intellectually exciting. I knew my mind was being stretched, and I loved the feeling. Yet I still felt out of place. Holy Cross had been a middle-class school full of strivers—people like me, even if most of them didn't look like me. Yale was different, and I felt the difference in my bones: I was among the elite, and I knew that no amount of striving would make me one of them.

The fact that I was black didn't enter into it at first. I thought of myself more as disadvantaged than as black, and I asked Yale to take that fact into account when I applied, not thinking that there might be anything wrong with doing so. I simply took it for granted that Yale was giving me a break because I was poor (and especially since that poverty was in part due to racial discrimination), in the same way that other students were given preference because they came from wealthy families or had parents who'd gone to Yale. I had been told that minority students were admitted under the same standards as these latter "legacy" students—and why shouldn't Yale be willing to take the same chance on a poor black kid from Georgia who'd always managed to achieve against the odds that it took on privileged white kids who had most of the advantages? I knew what I could do, and hoped to have another chance to prove it. I had spent the first part of my life fighting to overcome the disadvantages of my background, and I'd managed to excel at Holy Cross, Immaculate Conception, and Saint John Vianney, none of which then had special admission programs for blacks. I expected to do the same at Yale.

But in the years following Dr. King's assassination, affirmative action (though it wasn't yet called that) had become a fact of life at American colleges and universities, and before long I realized that those blacks who benefited from it were being judged by a double standard. As much as it stung to be told that I'd done well in the seminary *despite* my race, it was far worse to feel that I was now at Yale *because* of it. I sought to vanquish the perception that I was somehow inferior to my white classmates by obtaining special permission to carry more than the maximum number of credit hours and by taking a rigorous curriculum of courses in such traditional areas as corporate law, bankruptcy, and commercial transactions. How could anyone dare to doubt my abilities if I excelled in such demanding classes? I even went out of my way to take a course in taxation taught by Professor Boris Bitker after I heard from one black student that "Bitker flunks black students," and the honors grade he gave me—my first at Yale—would be my most satisfying experience in law school.* But it was futile for me to suppose that I could escape the stigmatizing effects of racial preference, and I began to fear that it would be used forever after to discount my achievements.

After graduating from Yale, I met a black alumnus of the University of Michigan Law School who told me that he'd made a point of not mentioning his race on his application. I wished with all my heart that I'd done the same. By then I knew I'd made a mistake in going to Yale. I felt as though I'd been tricked, that some of the people who claimed to be helping me were in fact hurting me. This knowledge caused the anger I thought I had put behind me at Holy Cross to flare up yet again, only in a different form. I was bitter toward the white bigots whom I held responsible for the unjust treatment of blacks, but even more bitter toward those ostensibly unprejudiced whites who pretended to side with black people while using them to further their own political and social ends, turning against them when it suited their purposes. At least southerners

* The professors at Yale Law School gave out four grades: fail, low pass, pass, and honors.

were up front about their bigotry: you knew exactly where they were coming from, just like the Georgia rattlesnakes that always let you know when they were ready to strike. Not so the paternalistic big-city whites who offered you a helping hand so long as you were careful to agree with them, but slapped you down if you started acting as if you didn't know your place. Like the water moccasin, they struck without warning—and now I had stepped within striking distance.

I began to consider the possibility of transferring to another law school, perhaps in the South. Then Kathy told me she was pregnant, and I knew I'd better stay put. We were already having trouble paying our bills, and I knew it would make no sense for me to add to our burden by switching schools. Too much work, too many doubts, not enough money, a baby on the way: I felt as though I was drowning in a swamp of troubles. I asked the financial-aid office if they could help, but I'd come at the wrong time of the year—their coffers were temporarily depleted—and so I applied for a number of university-wide scholarships. I was only invited to interview for one of them, a scholarship established by the Beinecke family, which had made a fortune from S&H Green Stamps. One of the Beineckes, I believe, interviewed me in the Beinecke Rare Book and Manuscript Library, which the family had donated to Yale, but his manner was cold and aloof, and even before we finished talking, I knew I didn't have a chance. I still wonder why a man of his wealth and standing felt the need to humiliate a desperate kid from Georgia over a few hundred dollars.

In the end all Yale had to offer me was the tuition postponement option, a program in which the cost of student loans was spread across a class of students who repaid it as a group according to their means, with the greatest burden falling on those with the largest incomes. I didn't know what else to do, so I signed on the dotted line, and spent the next two decades paying off the money I'd borrowed during my last two years at Yale. I was still making payments when I joined the

Supreme Court. One student pointed out that I was eligible for public assistance and food stamps, since I had little income and no support from my family. Another suggested that I declare bankruptcy after graduating in order to get out from under the crushing weight of all my student loans (a loophole that has since been plugged).

I may have been cynical, but at least I wasn't dishonest. I knew that the only way I'd be able to live with myself was to do as Daddy would have done. I thought of how he would point to the crops flourishing in the fields of our family farm and remind Myers and me that this was why we'd worked so hard; I remembered, too, how he liked to rattle the loose change in his pocket and tell us that the same coins had been in his pocket at the beginning of the week, and would still be there at the end. That was how I would have to live. I knew it was pointless to resent the good luck of my well-off classmates who I thought had it all—money, sophistication, contacts. Instead I had to find a way to draw positive energy from my situation, to keep my mind on the things I would be able to do in the future as long as I stayed out of the trap of spending money I didn't have on things I didn't need.

I kept telling myself that my problems were insignificant compared to the ones that Daddy and his generation had faced, and that my own lot had once been far harder. Myers and I had always wanted to be just like the other kids in Savannah, and couldn't understand why Daddy refused to give in to our childish desires. Why couldn't we watch TV or play with our friends instead of going out to Liberty County to pluck chickens and till the fields? Why did we have to go without underwear and decent shoes all summer? But Daddy knew it was more important for us to be financially independent than to do as the other kids did, and now, years later, I was profiting from his foresight: I was a college graduate who was about to earn an Ivy League law degree. This was no time to change my ways—especially now that I was about to become a father.

Kathy went into labor early on the morning of February 15, 1973, and our son was born that same evening at Yale New Haven Hospital. It had become fashionable for black parents to give their children African or Arabic names, so we called him Jamal Adeen Thomas. I'd been worrying myself sick over whether I was up to the demands of fatherhood and remained skeptical about the future of my marriage, but as soon as the doctor handed Jamal to me, I was filled with a profound sense of love and responsibility. Having been abandoned by my own father, I knew in my heart that I would do things differently, that I would gladly give up my own life to save the tiny boy I held in my hands.

Jamal's birth made me start thinking in personal terms about one of the biggest news stories of the year, the ongoing controversy over the use of busing to integrate Boston's public schools. I wasn't surprised by the explosion of white rage that threatened to rip the city in two. It was in Boston, not Georgia, that a white man had called me nigger for the first time. I'd already found New England to be far less honest about race than the South, and I bristled at the self-righteous sanctimony with which so many of the northerners at Yale glibly discussed the South's racial problems. Now that their own troubles were on national display, I was unsympathetic. The water moccasins, it seemed, were biting themselves.

None of this meant that I was in favor of busing, however. As I watched TV pictures of black children being bused into South Boston, it was clear that the situation had reached the point of total absurdity. *I* wouldn't have gone into South Boston. It would have been taking my life in my hands for me to do so. Why, then, were innocent children being made to do what a grown man feared—and to what end? Aside from the threat of violence, the white schools in South Boston were at least as bad as the ones in black neighborhoods, so what was the point of shipping those children from one rotten school to another? I'd long believed that the best thing to do was to stop government-sanctioned segregation, then concentrate on education and equal employment opportunities. The rest, I thought, would take care of itself. But once again blacks were being offered up as human sacrifices to the great god of theory, and I swore on the spot never to let Jamal go to a public school, even if I had to starve to pay his tuition. I had no intention of allowing my son to become a guinea pig in some harebrained social experiment.

In the seventies you rarely had to look very far to find a theory, or a black person on whom it was being tried out. Like every other black law student, I was uncomfortably aware that blacks failed to pass the bar exams at a much higher rate than whites, and that the NAACP Legal Defense and Education Fund had filed lawsuits alleging that the exams they took were racially discriminatory. Lani Guinier, one of my classmates, was involved with the Legal Defense Fund, so I asked her to supply me with information about the extent of the problem and the strategy that the Legal Defense Fund was pursuing. At first I assumed that the disproportionate black failure rate was conclusive evidence of racial discrimination, but the more closely I looked at the facts, the more apparent it became that I was wrong. At that time each question on the bar exam was graded separately by a different scorer and each completed exam identified solely by number, thus making it impossible for the graders to tell which examinees, if any, were black. Some claimed that blacks wrote in "black English" and thus could be identified from the syntax of their responses, but in addition to finding that unlikely, I didn't think it unfair to expect lawyers representing their clients in a court of law to be able to write in standard English. In any case the inability of a black law student to write and speak English properly wasn't evidence of discrimination by the graders—it was an indictment of the quality of the education he had received. This left only one argument, the Legal Defense Fund's "adverse impact" theory, which held that if a neutral examination produced disparate results among the races, then it could be considered discriminatory. But I didn't buy that, either, knowing that no measurement of any part of our lives ever produced identical results for all racial or ethnic groups. To argue otherwise, I thought, diverted attention from the real culprits, the people who were responsible for the useless education these young people had received.

The problem with my analysis, of course, was that it was of no help to those black students who had already finished school and now found themselves unable to pass the bar exam. But the adverse-impact theory had its own built-in problem, which was that its advocates appeared to be suggesting, knowingly or not, that blacks could never catch up with whites. Neither alternative was attractive to me, and I had no easy solution of my own to offer, but at least I'd thought the problem through for myself instead of jumping to a quick and easy conclusion that might be emotionally satisfying but failed to fit the facts. This, I decided, was the right way to approach any problem that excited my passions, and if it led me to disagree with the solutions that were generally accepted, or to advocate positions that would make me unpopular—especially when it came to matters of race—then so be it.

LAW STUDENTS USUALLY spend the summer in between their second and third years working for a firm they hope will offer them a full-time job once they graduate. Many of my Yale classmates hoped to land jobs at prestigious firms in New York or Washington, D.C., but I still wanted to go back to Savannah, so I spent the summer working for Bobby Hill, a member of the Georgia legislature who was also one of Savannah's top black attorneys. Bobby (as everybody called him) was flamboyant, brilliant, and courageous. I told Lani Guinier that I hoped to return to Savannah after graduating from Yale and work for a black law firm that handled civil-rights cases, and she helped me obtain a $60-a-week Law Students Civil Rights Research Council grant from the Legal Defense Fund to spend the summer of 1973 working at Hill, Jones, and Farrington, Bobby's firm. When Bobby offered to pay me an extra $40 a week on top of that, I accepted at once. It was the only summer job for which I applied, and the only one I wanted. A hundred dollars a week wouldn't go very far, but Daddy told me that he was willing to let Kathy, Jamal, and me live in his Savannah house while he and Aunt Tina spent the summer at the farm, and I figured that if we were careful, we could make ends meet.

In May Kathy and I loaded up our not-so-trusty Volvo and headed South with Jamal. Foreign cars were still relatively uncommon and ours was old and worn, so I had it inspected and took along a tool kit and plenty of spare parts. The prospect of driving through the South frightened me, with good reason. One of the first things we saw as we drove across the state line into North Carolina on Highway 301 was a billboard that said: The United Klans of America Welcome You. Not long after that, our engine started to make a loud, clattering noise that wouldn't go away. At sunset I found a small garage close to the highway. All of the mechanics there were white, and they stared at us strangely. One of them came out to inspect the Volvo, which he referred to as "one o' them foreign cars." He looked the engine over thoroughly, told us that there was no one in the area who knew how to fix it, then suggested that we find a place to spend the night and see if we could make it to a bigger town the next day. I was sick with worry, but I found a motel that would put us up, and after a bad night's sleep we set out once more. The noise coming from the engine grew steadily louder, and early that afternoon, it finally sputtered and failed. Fortunately we had reached Orangeburg, a small South Carolina town, and I pulled off the road into a gas station that turned out to be run by a black man. He, too, was unable to fix the car, so I called Myers, who now lived in Savannah, and he agreed to drive up to Orangeburg and take us back home. We left the Volvo at the gas station, and I ordered the necessary parts with money from my first paycheck, returning several weeks later to repair and retrieve the car. It was an inauspicious start to a frustrating summer.

My first day at Hill, Jones, and Farrington was an eye-opener. Bobby Hill was a sharp dresser who drove a brand-new dark blue Lincoln Continental and exuded confidence as he strolled into the office late each morning, invariably keeping his first client of the day waiting. After he introduced me to the staff, I was escorted to the small law library on the second floor of the refurbished Victorian house that served as the firm's Savannah office. I looked through the French doors at the quiet street below, lined with large oak trees, occasional magnolia trees, and lush azalea bushes, and thought my dream had come true at last. I couldn't wait to start putting in long hours, seven days a week, to right the racial wrongs of my hometown.

It didn't take long for me to see that Hill, Jones, and Farrington wasn't a seven-day-a-week firm. Bobby and his partner Fletcher Farrington worked on separate projects, meaning that there were no cases on which the entire staff collaborated. Fletcher, who was white, had worked in the Justice Department and was a native of Alabama, and though I wasn't privy to the firm's finances, I got the definite impression that it relied heavily on some of his large lawsuits (as well as another large suit that the Atlanta office was handling) to cover its operating budget. As for Bobby, his casual demeanor seemed to inspire others to come and go as they pleased and do their work lackadaisically; we ate a lot of leisurely lunches at a nearby restaurant, and late-afternoon drinking was commonplace. At first it was fun listening to Bobby tell stories about his career, but I would have preferred more conversations about the law. It wasn't unusual for some of my coworkers to still be drinking when I left for the day, and more than a few of their bull sessions continued late into the evening. Each day I showed up for work bright and early, hoping for a big project that would allow me to learn more about the practice of law in Savannah, but I spent most of my time sitting in the upstairs library, impatiently twiddling my thumbs.

In an attempt to keep me busy, Bobby asked me to participate in a case involving a black family that owned land in Hilton Head, South Carolina. Today Hilton Head is a well-established island resort, but in the early seventies, developers were still voraciously buying up land on which to build the resorts and golf courses that have since made it a popular destination for wealthy tourists, and they weren't always scrupulous about how they got their hands on it. A middle-aged black man whose family owned land on Hilton Head came to Bobby to ask for help in fighting off an attempt by the local taxing authorities to seize his property to settle a delinquent-tax assessment. It turned out that the taxes had been paid all along by the man's grandmother, who had kept the receipts tucked away in her Bible. (It is not unusual for older southern blacks to file important papers in their family Bibles.) Once she produced the receipts, the developers offered her a few thousand dollars for the small parcel of land, which was in the middle of a planned golf course. Bobby asked me what I thought the family should do; I knew a thing or two about hogs, so I suggested that they fence in the land and populate it with the foulest-smelling swine they could find. Needless to say, Bobby chose instead to extract as much money as he could out of the developers, who ended up paying the family some $60,000. It was a fair enough price, but the sharp practice of the developers stuck in my craw, and for more than three decades, I made a point of steering clear of Hilton Head.

The most exciting case I worked on was that of Carl Isaacs, a nineteen-year-old prisoner who escaped from a Maryland jail and stole his way south to Donalsonville, a little town in southwest

Georgia, accompanied by his brother, his half-brother, and a friend. There they broke into a mobile home, and as the members of the family living there returned home that day, Isaacs and his companions murdered them one by one, raping one woman repeatedly before shooting her dead. By the time Bobby and I entered the case, Isaacs had been captured and charged with capital murder, and we flew over to Donalsonville to represent him. When we got there, we found a small crowd of white people milling about in front of the courthouse. I wondered if they might be a lynch mob, but Bobby didn't turn a hair. Everyone inside was white as well—it looked just like the courtroom scene in *To Kill a Mockingbird*—and they all glared as we walked in. Then it hit us: we were late. The closest airport to Donalsonville was in Alabama, and when Bobby and I had flown there the night before, we'd automatically set our watches back an hour, forgetting that Donalsonville was on the other side of the time-zone line.

As Bobby coolly explained our mistake to the judge, I sat paralyzed with fear, trying to imagine what the stern-faced white people sitting all around us might be inclined to do to a pair of uppity black lawyers representing a man who had slaughtered a white family and raped a white woman. Then Isaacs was brought out, handcuffed and bedecked in bright prison garb. It was the first time we had seen him—and he was white. I've never been more relieved to see a white man in my life. I returned home full of admiration for Bobby's unflappable calm. I didn't think much of the way he ran his office, but I envied his nerve.*

Kathy and I pinched our pennies that summer, shopping with a small plastic clicker in order to keep a close watch on our grocery expenses and always making sure that Jamal's needs were met before

---

* Isaacs was duly convicted, but managed to put off his fate for three decades, and early in my tenure on the Supreme Court, he filed a motion asking me to recuse myself from his case. He was finally executed by lethal injection in 2003.

we spent anything on ourselves. It was disconcerting to watch other people using food stamps to buy whatever they pleased, but I knew our financial problems would someday come to an end, whereas theirs were likely to stay with them. It was a pleasure to watch Jamal grow up, and to see how much Daddy enjoyed being with him. It pleased me, too, that Daddy and I were feeling somewhat more comfortable with each other, perhaps because I had made a point of telling him that I planned to come back to Savannah to practice law. He didn't think much of my new ideas about America, which now vacillated unpredictably between cynicism, radicalism, and patriotism, and told me for the umpteenth time that Yale had turned me into an "educated fool." But our relationship had become easier all the same, and I was glad for that.

Myers was also living with us in Daddy's house that summer, carrying a full load of courses at Savannah State and working every night at a local motel. I couldn't see how he managed it, but he never complained. "I got it from Daddy," he said. He had just come back from a tour of duty in the Air Force, and the experience had made him every bit as patriotic as Daddy. He chided me about my radical views, asking why I didn't leave the country if I thought things were so bad here. The thought had never occurred to me, even when I heard my black contemporaries proclaiming the virtues of returning to Africa. My roots were in Georgia, not Africa, and I had every intention of going there as soon as I got my degree. For all my cynicism, I still found it impossible to hate my own country, and I'd found it deeply disturbing when Jane Fonda had criticized America during a visit to North Vietnam the preceding year.

Myers and I hadn't spent much time together since Daddy had kicked me out of the house in the summer of 1968. After that we'd spent a few weeks living in Leola's apartment while he was waiting to enter the Air Force. It was a hard time for both of us. Myers was full of bitterness toward Daddy and Sister Mary Virgilius, the eighth-grade teacher and principal of St. Benedict's, who'd made him repeat the eighth grade. He swore never to return to Savannah, but now we were together again. We reconnected that summer, and stayed close throughout the rest of his too-short life.

At summer's end Bobby and Fletcher asked if I wanted to work for them permanently after graduating from Yale. I told them that while I hadn't yet made up my mind for sure, I thought I'd probably look elsewhere. The truth was that I'd seen enough to know that Hill, Jones, and Farrington wasn't my kind of firm. I was deadly serious about the law, and I knew that working with Bobby wouldn't be sufficiently challenging to satisfy me. My instincts were sound: Fletcher left the firm four years later, while Bobby and Clarence Martin, one of the junior partners, were subsequently disbarred and convicted of a variety of crimes. But all this was in the future. What mattered now was finding another job as quickly as possible, and so I spent the first semester of my third year at Yale researching law firms. Until then I'd never considered any big-city firms—I'd taken it for granted that I was going back to Savannah—and I didn't like the idea any better now. I had grave reservations about working at a predominantly white institution, subject to the whims of white superiors. But it was starting to look as if I might not have any alternative, so I got a credit card from the small Saks Fifth Avenue store near the law school, used it to purchase the first new suit I had owned since childhood, and started looking for a place to work.

I set my sights on Atlanta, but I also interviewed with firms in New York, Washington, and Los Angeles that were recommended by my classmates, who assured me that I'd have no difficulty finding a job. They were wrong. One high-priced lawyer after another treated me dismissively, making it clear that they had no interest in me despite my Ivy League pedigree. Many asked pointed questions unsubtly suggesting that they doubted I was as smart as my grades indicated. A firm in Atlanta briefly seemed interested in me, then started to blow hot and cold, stringing me along with expressions of interest but refusing to make a commitment. When they hired one of my classmates, I decided I'd had enough: I called them up and told them I wasn't interested anymore. By late December I had yet to receive a single job offer. Now I knew what a law degree from Yale was worth when it bore the taint of racial preference. I was humiliated—and desperate. The snake had struck.

I mentioned my dire situation to Frank Washington, a classmate and friend, who told me that John Danforth, a Yale Law School graduate who was now serving as Missouri's attorney general, was looking for other Yalies to work for him. I quickly made an appointment to see the attorney general, who was coming to New Haven in a few weeks. I showed up early and killed time by pacing nervously outside the house where he was staying. I knocked on the door at the appointed hour, but no one answered. Had he forgotten about me? I knocked several more times. Just as I was getting ready to leave, the door was opened by a tall, imposing man with a booming voice. His dark head of hair was lightened by a single shock of white. "How are ya?" he asked before I could say a word. I knew at once that this interview was going to be different.

"Clarence, there's plenty of room at the top," the attorney general said as we sat down to talk. *That's easy for you to say*, I thought, knowing that he was one of the heirs to the Ralston Purina fortune and had been elected attorney general of Missouri while he was still in his early thirties. Maybe there was room at the top for people like him, but so far I hadn't even managed to find it at the bottom. He seemed to read my thoughts, for the next thing he said was that since he was white and had been born with a "platinum spoon" in his mouth, he couldn't begin to imagine what my life had been like. That disarmed me, and the longer we talked, the surer I became that he was as genuine as he seemed to be. After what I'd been through, his honesty and sincerity meant almost as much to me as a paying job.

Most of the lawyers with whom I'd interviewed had made a point of telling me how lucky I'd be if they offered me a job. Not Jack Danforth. He told me how he wanted to change things in Missouri. To do that, he said, he needed talented young assistants, not just from his own state but from all over America. The starting salary, he admitted, wasn't very impressive—$10,800 a year—and I didn't see how I could afford to move to Jefferson City and pay my bills for that kind of money. But I was intrigued, and asked him point blank if he would treat me the same as everyone else in the office, no better and no worse. He said he would.

Shortly after the interview, I received a letter from the attorney general offering me a job. He also offered to fly Kathy and me to Jefferson City, the state capital of Missouri, so that we could see it for ourselves. We took him up on his offer a couple of weeks later, traveling there on a dreary, foggy day. We went straight from the airport to the Missouri Supreme Court building, where we ate a box lunch and chatted with several staff lawyers, all of whom seemed bright, energetic, and unpretentious. I was interviewing them as much as they were interviewing me, and I returned to New Haven sure that I wanted to work with them—and with their boss. I still had my doubts about the salary, and I was also dubious about working for a Republican, as were a number of my liberal classmates. But I assured them that he seemed to be a good man in spite of his politics, and most of them thought it would be acceptable for me to hold my nose and take the job. "What a waste of a Yale Law School education!" one of them blurted out. "I'll see you in twenty years," I retorted, knowing that it wasn't much of a comeback.

I accepted the offer in January. I had to pass the Missouri bar exam, so I arranged to take a bar-review course in St. Louis that coming summer. I asked the attorney general if he could help me find a place to stay, telling him frankly that I was all but broke, and he put me in touch with Margaret Bush Wilson, one of St. Louis's most successful black lawyers and best-known civic leaders. She told me that her son Robert had decided to spend the summer away from home and that I could fill in for him.

Now that I had a job lined up, the rest of the semester flew by. At last it seemed that things were going my way, though fate had one more piece of bad news in store for me. I asked Daddy and Aunt Tina to come to my graduation ceremony, and Aunt Tina sent me back a letter saying that they'd both had a bad year, medically as well as financially, and couldn't come. Later on Daddy explained that he had to stay home to take care of Cousin Robert Chip, who was dying of prostate cancer. He added that he also had to plant his crops and couldn't spare the time for the long trip north. Once again I offered to pay for a plane ticket to New Haven so that he could return home immediately after the ceremony, once again he turned me down, and once again I was crushed. Without my grandparents, especially Daddy, the graduation ceremony would be just that—a ceremony, nothing more.

I graduated from Yale Law School on May 20, 1974. Kathy and her parents were in the audience, along with one of my cousins who lived in Washington. I mostly felt relieved, as well as a little bit lonely. I'd spent twenty hard years in school, and now I was done. Only the bar exam stood between me and my law license. Attorney General Danforth told me that he had yet to hire anyone who failed the bar exam. I was determined not to be the first.

5

THE GOLDEN HANDCUFFS

Kathy and I stored our furniture in the barn behind her parents' house, where she and Jamal would stay while I went to St. Louis to prepare for the bar exam. First, though, I had to go to Savannah to pick up our old car, which we'd left in Daddy's garage the preceding summer, fearing that it wouldn't survive the trip back to New Haven. I'd bought yet another used Volvo while we were down south, an oil-burning clunker that wasn't worth the $800 I paid for it. I didn't like the idea of driving our older Volvo all the way to St. Louis, but I couldn't see any alternative. My brief reunion with Daddy, not surprisingly, was uncomfortable: he couldn't understand why I wouldn't go to work for Bobby Hill, and I couldn't understand why he hadn't come to New Haven to see me graduate. Again we found ourselves at odds.

When I finally made it to St. Louis, I was surprised to find that Margaret Bush Wilson lived in the heart of the city, in a once-fashionable neighborhood ringed by urban decay. She had grown up in the brick row house that she now shared with her older brother James and her ex-husband, Robert Wilson. She greeted me with open arms and put me in her own second-floor bedroom, moving across the hall to a smaller room. I soon discovered that the members of my new family were well read and very opinionated and liked nothing better than a good scrap. Not long after I arrived, I made the mistake of saying that I thought economically disadvantaged students should be the chief beneficiaries of preferential admissions policies at colleges and law schools, since they were the ones who had suffered most from past discrimination and so deserved the greatest amount of help. Mrs. Wilson disagreed politely but firmly, telling me in no uncertain terms that the children of middle- and upper-class families would find it far easier to bridge the cultural gap and thus help to break down racial barriers. "They know which fork to use," she told me. I couldn't see how growing up in a one-fork family had hurt me, but I didn't go out of my way to offend her, either. Mostly I stayed out of the family debates, preferring to listen and learn.

Mrs. Wilson and her unconventional family had a collective passion for reading that I found inspiring. (It also impressed me to no end to watch Mr. Wilson open up the paper each morning and breeze through the daily crossword puzzle—using a pen.) Not having grown up in a bookish household, I loved being around older black people who enjoyed talking about literature and ideas. For the moment, though, I had to stick to my studies. I arose by six each morning to get in a half hour of exercise before studying, then walking to St. Louis University Law School, where I studied the rest of the day in preparation for my nightly bar-review classes. I didn't have much time to spend money, but I didn't have much money to spend, either, and by the middle of June I was flat broke. I tried to sell my blood—a local blood bank paid twenty dollars a pint—but my pulse rate was too low and the nurse turned me down. I glanced around the room, noticed that most of the other donors were booze-soaked bums, and begged her to try again. The result was the same. I trudged home, thinking to myself that a man who couldn't even sell his own blood to buy a decent meal had sunk pretty low.

Finally I called the attorney general to ask for help, and Alex Netchvolodoff, his administrative assistant, arranged for me to do some part-time work in the St. Louis office. I couldn't afford to ride the bus there, so I walked to and from the office, which took about an hour. That was the summer of Richard Nixon's impeachment hearings, but I didn't follow them very closely. I was too busy worrying about my own problems to concern myself with the fate of a Republican president for whom I hadn't voted. One thing I'd learned at Yale was how to study for a tough exam: John Bolton had taught me the secret of distilling all the material in a course into a succession of shorter and shorter outlines, ending up with a concentrated super-outline that fit on a single index card. John's study method would serve me well many times, but I gave it the workout of my life in the summer of 1974.

I went to Jefferson City at the end of July to take the two-day exam. It went smoothly, and I returned to St. Louis concerned about the outcome but sure I'd done my best. I was sorry to say good-bye to Mrs. Wilson and her family; I knew they'd given me something precious and irreplaceable that summer. I timidly asked how much rent I owed her. "You don't owe me a thing," she said, explaining that I'd more than repaid her by filling the void left by the absence of her son. "All I ask is that someday you do for someone else what I've done for you."

A few days after I returned to Worcester, Kathy, Jamal, and I set out on the thousand-mile drive to our new home, a small but comfortable apartment in Jefferson City that I'd found earlier in the summer. Kathy had decided to go back to college, and our apartment was close to Lincoln University, the historically black school she would be attending. I reported for work at the Missouri Supreme Court building the day after we arrived and was ushered downstairs to the part of the basement where the criminal-appeals attorneys worked. Preston Dean, who ran the division, showed me to my tiny, windowless office, handed me a stack of old briefs, and told me to spend the rest of the week reading through them in order to familiarize myself with the process. A short while later he brought in an even larger stack of new briefs and trial transcripts, said they were mine, and told me to get to work on them right away. "What happened to that week I was supposed to spend getting myself up to speed?" I asked. "It's been compressed," he said.

I sensed early on that some of my new colleagues were concerned about my attitude—or, rather, their assumptions about my attitude. Not only did they expect me to have a chip on my shoulder because I was black, but some of them also expected me to be a self-absorbed Ivy League elitist. (That struck me as funny, considering how I felt about Yale.) But once they saw that I wasn't what they'd been expecting, they loosened up—and so did I. It was true that I was the only black attorney in the office, but that didn't make me the resident expert on race relations, nor did it mean that I had to be handled with kid gloves. I didn't represent all blacks any more than they represented all whites. I was just another young attorney trying to do a good job. That was all I'd ever wanted to be at Yale, but race kept getting in the way; in Jefferson City, by contrast, I was treated like everyone else, and I appreciated it.

Our office was grossly understaffed. That came as no surprise to me, since the attorney general had promised at our first interview that he could give me more work for less money than anyone else in the country. (I'd thought he was joking.) What bothered me more was that as a criminal-appeals attorney, I would have to argue in favor of keeping blacks in jail. I still thought of most imprisoned blacks as political prisoners. I had no facts to back up this opinion, a reflex response left over from my radical days, and didn't need any: I knew that anything "the man" did to black people was oppression, pure and simple. What changed my mind was the case of a black man convicted of raping and sodomizing a black woman in Kansas City after holding a sharp can opener at the throat of her small son. He was no political prisoner—he was a vicious thug. Perhaps he and the woman he'd brutalized had both been victims of racism, but if that were so, then she'd been victimized twice, first by "the man" and then by the thug. This case, I later learned, was far from unusual: it turned out that blacks were responsible for almost 80 percent of violent crimes committed against blacks, and killed over 90 percent of black murder victims. This was a bitter pill to swallow. Until then I'd ignored the obvious implications of black-on-black crime rates. After I worked on that case, I knew better than to assume that whites were responsible for all the woes of blacks, and stopped throwing around the word "oppression" so carelessly. I also grew more wary of unsupported generalizations and conspiracy theories, both of which had become indispensable features of radical argument.

I'd had the bar exam on my mind ever since I reported for work, and in early September, Tom Simon, the Supreme Court clerk, told all the newly hired attorneys in the attorney general's office that the results would be announced that evening. The custom was for us to gather in his office to have a beer as we waited for the posting of the results shortly after midnight. Tom did what he could to calm us down, but my nerves were shot, in part because one of the blacks in my bar-review course had warned me that Missouri had a "rule of two," meaning that only two blacks were allowed to pass each year. I didn't take what he said seriously, but so much depended on my passing the bar that I found his warning impossible to ignore. As we drank our beers and watched the clock, I couldn't decide whether to wish that midnight would come sooner or that it would never come at all. Finally Tom unlocked his office safe, took out the envelope containing the results, and opened it slowly and deliberately. I felt like throwing up. "All of you passed," he said. The relief that flooded over me was overwhelming. I called Kathy at once—but not Savannah. Daddy, I felt, had burned that bridge.

I was sworn in as a member of the Missouri bar on September 14, 1974. Preston Dean congratulated me when I reported for work the following Monday, then told me I'd been assigned to argue a case before the Missouri Supreme Court the very next morning. One of the older attorneys, he explained, would introduce me to the court, but after that I was on my own. *You've got to be kidding,* I thought. As I read the briefs and trial transcript, I remembered the stories I'd heard about how many new attorneys became physically ill before their first argument. Some, it was said, simply couldn't go through with it. I didn't get much sleep that night.

My legs were rubbery and my stomach queasy as I climbed the stairs to the Supreme Court on Tuesday, accompanied by Neil MacFarlane, one of the criminal division's senior attorneys. For a moment I wanted to run away. Wild thoughts raced through my head when we entered the handsome courtroom and Neil introduced me to the judges. How would I react under fire? Would I lose control? Did it matter that I was black? Couldn't someone else argue this case? Neil wished me luck and left. The attorney for the appellant was wearing a burnt orange polyester suit with matching shirt and tie, and he waved his arms wildly as he made his impassioned argument. The judges seemed annoyed by his manner—not to mention his outfit. I was glad I'd worn the blue pinstripe interview suit I'd bought in New Haven with my Saks Fifth Avenue card, especially since I owned an even gaudier thirty-five-dollar suit of my own, a blue coat with red trim that came with two pairs of pants, one red with blue trim and the other blue with red trim. I said my piece and sat down quickly, and the judges asked me no questions. Breathing a sigh of relief that I'd left my thirty-five-dollar suit in the closet that day, I made a mental note to avoid histrionics and stick to pinstripes and white shirts.

That first day on the job hadn't been a fluke: Preston would simply walk into our offices and drop large stacks of cases on our desks. We usually had no more than a month or so to prepare each brief, so I tried to start reading the new case materials as soon as I got them. I read the appellant's brief first, then the transcripts, making detailed notes on a legal pad. One of the older attorneys in the office had told me that while it was sometimes excusable not to know all of the law, there was never any excuse for not knowing the facts. Since the transcript was the trial record, it was important to get it under my belt right away. Once that was done, I went to our small law library or walked upstairs to the Supreme Court library to do the necessary research. After reading the relevant cases, I began drafting my brief, always starting with an outline. In a simple case, this would take only a day or two; in more complicated cases, I might spread the work out over several weeks, setting aside the project at intervals to do other tasks. Once a draft brief was completed, the practice was to circulate it for peer review in order to have another set of eyes look for mistakes or opportunities for improvement before we sent the brief off to the printer. Multiply that process countless times and you'll get a sense of how I spent my first months in Jefferson City.

I grew more confident of my ability to cope with the crushing workload, and it wasn't long before I started making new friends. One of them, Richard Wieler, was a quadriplegic who had contracted polio at the age of fifteen. His parents were told that he'd be dead before he was twenty-one. Instead he graduated with honors from the University of Missouri Law School. Dick could move only one finger: his hands were always perched on a pair of metal armrests built into his wheelchair, and he used a pointer attached to a mouthpiece to move papers, turn pages in a book, and write. I'd never known anyone who was so severely handicapped, and at first I found it uncomfortable to be around him. But then I stopped thinking of Dick as a "cripple" (a word that quickly vanished from my vocabulary) and started to see him as the kind, courageous human being he really was. Though his physical condition was impossible to overlook, it didn't define him any more than the color of my skin defined me. Getting to know Dick Wieler taught me a lesson I never forgot, though it took a long time for me to absorb it completely. How could I possibly allow myself to be consumed by anger after seeing the grace and dignity with which he confronted his own misfortune?

Jefferson City was a good place to reflect on such things, for it was a sleepy little town where nothing much ever happened. I remember opening the local paper one morning and noting with amusement that the big story of the day was that someone had been turning over trash cans. The peaceful atmosphere was just what I needed after a decade of personal turmoil. So was the fact that most of my new friends and neighbors appeared to have no ideological axes to grind: they argued about sports, not politics. Not that they didn't have political opinions, too, but their points of view ranged all the way across the spectrum, with a generous sprinkling of indifference in between. As a result there was none of the herd mentality that I'd seen in New England. I could stake out a dissenting position without having to worry about giving offense, and I no longer needed to be on guard against racial slights. Everyone in the office came in for his fair share of kidding; everyone was as honest with me as I was with them. I joined a weekly basketball game at the highway patrol gym and was delighted by the easygoing atmosphere. I was also relieved to find that none of the white secretaries with whom I worked subscribed to the radical political views to which Yale had accustomed me. Many of the women I'd met there had come from the most privileged of circumstances, yet they often referred to themselves as "oppressed." I found it hard to take their "oppression" seriously, since I'd spent the first part of my life living among black women who cooked and kept house for the middle- and upper-class whites of Savannah. *They* never talked about being oppressed. What right, then, did the elite white women of Yale have to complain about their lot? I had a feeling that the decent, hardworking women in the attorney general's office, most of whom came from the small towns and farms around Jefferson City, would have felt the same way I did.

Before long I began to relax, and to see and live life more fully. I'd been spending so much time thinking obsessively about race that I'd lost sight of the rest of what the world had to offer. My new friends knew better. They understood what mattered: family, home, church, friends. Once I had cynically supposed that no one who lived like that could be truly happy or sincere, but now I was surrounded by people who were both. It felt like home.

I FINISHED MY stint in the criminal-appeals division in February. Now it was time to transfer to the civil division, and though just one position remained open—representing the Department of Revenue and the State Tax Commission—I liked the sound of it. Not only was I interested in tax and corporate law, but the job would allow me to work independently on important cases, going head to head with some of Missouri's finest lawyers. That was exactly what I wanted. I knew the only way I could become as good as they were was to litigate against them, and I also knew that doing so would boost my professional credibility. I meant to be employable when I left Jefferson City. I'd learned the hard way that a law degree from Yale meant one thing for white graduates and another for blacks, no matter how much anyone denied it; I couldn't do anything about that now, but I had a feeling that winning real cases in court would be a better demonstration of what I could do than a law school transcript. As a symbol of my disillusionment, I peeled a fifteen-cent price sticker off a package of cigars and stuck it on the frame of my law degree to remind myself of the mistake I'd made by going to Yale. I never did change my mind about its value. Instead of hanging it on the wall of my Supreme Court office, I stored it in the basement of my Virginia home—with the sticker still on the frame.

Walt Nowotny, my new supervisor, mostly left me on my own. At first I found my freedom daunting, but I grew to love it. Daddy had often told me that he could never work for another man, and I didn't like being told what to do any more than he had. From then on I found close supervision hard to take, becoming impatient whenever I was told to do things I thought trivial or unnecessary. This impatience would soon have a profound effect on the course of my career, though I didn't know it at the time. All I knew was that I appreciated not being overmanaged or forced to do irrelevant make-work in order to keep an overbearing boss off my back.

The only thing I didn't like about my new job was the pay. After taxes I took home about $560 a month. Kathy did her best to stretch what I made as far as it would go, but once we'd paid our monthly bills, we usually had less than $10 left to tide us over until the next paycheck, and I often found myself forced to borrow small sums from my colleagues in order to make ends meet. Dick Wieler in particular was always good for a five-dollar loan that I scrupulously repaid each month, only to borrow it again the next. Kathy and I were so strapped for cash that we washed our clothes by hand, just as Leola and Sister Annie had done when I was a little boy in Pinpoint. Sometimes we saved our nickels and took our wet clothes to the neighborhood laundromat to dry, but more often we made do with the backyard clothesline. I owned only two dress shirts, and I washed one or the other of them each night and ironed it the next morning before putting it on and going to work.

As I walked to the office one morning, I looked down and noticed a black wallet on the sidewalk. I picked it up, put it in my coat pocket, and continued on to the office. I opened the wallet to examine the contents as soon as I got there. I stopped counting the cash at around six hundred dollars. My first thought was that this must be a gift from heaven, but I quickly dismissed that foolish notion. I called the owner to tell him to come pick it up, and he spoke gruffly to me, not at all in the way you'd expect from a man whose wallet had been found. He came by the office later that morning, and turned out to be an older white man with a permanent scowl etched on his face. I handed him the wallet and he immediately started counting his money; I told him with audible irritation that it was all there. He stopped counting, fished out a five-dollar bill, and grudgingly handed it toward me. I refused it, saying that I hoped he'd return somebody else's wallet someday. He turned on his heel and stalked out of the office without saying another word. I learned two lessons that morning. The first one was that honesty is what you do when no one is looking. The second one was more important, so much so that I came to think of it as a defining moment in my ethical development: my needs, however great they might be, didn't convert wrong to right or bad to good. That man's wallet wasn't mine, no matter how much I needed the money, or how rude he happened to be. I often had occasion to remind myself in years to come that self-interest isn't a principle—it's just self-interest.

Every once in a while I managed to squeeze a laugh out of our near poverty. One day I went for a walk with Attorney General Danforth, who saw a basket of apples in a downtown grocery store window and asked me to loan him a quarter so that he could buy one. When I reluctantly admitted that I didn't have a quarter to spare, he pulled a twenty-dollar bill out of his pocket and bought one for each of us. But our financial situation was no laughing matter, and it became deadly serious when a bank foreclosed on one of my student loans. The bank, I discovered, had been sending its payment notices to my grandparents' house in Savannah. Not knowing what they were, Daddy and Aunt Tina had put them in a drawer for me to pick up on my next visit home. (They'd done the same thing with my grade reports.) Nobody at the bank had time to listen to me, so I called the regional office of the Department of Health, Education, and Welfare, which supervised the student-loan program. The man to whom I spoke suggested that I try to get a consumer loan to repay the bank, but because of my low salary and lack of credit history, I assumed that no one would be willing to take a chance on me. Once again the attorney general saved the day: I mentioned my problem to him, and he referred me to the president of a local bank, a friendly small-town type who believed that character matters as much as collateral. He took Jack Danforth's word for my character and agreed to lend me the money. "Don't disappoint me," he said as I left his office. I didn't.

The only good thing about not having any money to spare was that it helped to keep my consumption of alcohol under control. I went to bars only once or twice during my years at Holy Cross and Yale—I did my drinking at home—but in Jefferson City I felt so completely at ease with my new colleagues that I started going out with them after work from time to time. Since I had so little money, one of them always picked up the tab. Years later one of my drinking buddies mentioned that he'd never seen me pick up a tab in all the time we spent together, and I reluctantly admitted that he was right. It wasn't until I started earning more money that my drinking became troublesome, though even then it must have been an outward sign of the anxieties that simmered inside me.

After I'd spent a few months in the civil division, I was joined by two assistant attorneys general, Doug Mooney and John Ashcroft. Doug was a recent graduate of Washington University Law School.

John had previously been appointed by Kit Bond, the governor of Missouri, to fill the job of state auditor, an office Bond had vacated after winning the governorship. John later ran unsuccessfully for the position in an off-year election, and we all knew that he would someday run for another statewide office. John was a pleasant, devoutly religious man who made a sincere effort to get along with me. It bothered me, though, that he was one of the highest-paid assistants in the office while I was floundering financially, and I treated him snidely for no reason other than envy. John invited me to his house many times, but I always replied that I wouldn't come unless I could bring a six-pack with me, knowing full well that as a member of the Assembly of God Church, he could not allow drinking in his home. John was unfazed by my immature behavior and wholly Christian in his conduct toward me. He was as self-deprecating about his legal abilities as he was steadfast in his faith. Even now I regret the way I acted toward him.

The only other attorney who worked directly with me in the civil division was a young man named Joel Wilson. I met Joel when he was still a law student, spending the summer of 1975 as an intern in the attorney general's office. I noticed him wandering around the office and asked what he was doing. "Filing and copying," he said. I told him I had something more substantial for him to do and invited him to help me with one of my cases. His work was so outstanding that I let him argue the case in the Supreme Court of Missouri during his last year of law school. Joel joined the attorney general's office after graduation, and we became good friends. I wanted him to understand what life in the Deep South was like, so on one of my trips to Savannah I brought back a miniature Georgia flag—the same one that had been adopted in 1956, with the Confederate flag and the Georgia state seal displayed side by side—and asked him to try to imagine how he would have felt growing up under a flag like that had he been black. Joel was from the Bootheel section of southeast Missouri, where a black man had been lynched in 1942 and segregation remained in force well into the sixties. He got the point.

I had come to feel that virtually every discussion of race in America was fundamentally dishonest. The continuing controversy over busing was a prime example. Many whites were crying foul about school busing, saying that they wanted "neighborhood" schools for their children. This, of course, was nothing more than what blacks had wanted during segregation—the chance to attend good schools near their homes instead of being carted across town to all-black schools. Why, then, should they now wish to be carted across town to all-white schools? Daddy and I were at loggerheads about most things, but we found ourselves in full agreement when it came to busing. Even in a segregated world, education was our sole road to true independence, and what mattered most was the quality of the education that black children received, not the color of the students sitting next to them. "Nobody ever learned anything on a bus," Daddy said. Nor did either of us believe that busing would somehow help to ease white racism. What made anyone think that white people who didn't want to eat with us were going to like our children any better? We both saw busing as a tragic digression from the quest for real equality: the ostensible means of achieving a desirable end had become the end itself.

I felt sure that Daddy and I weren't the only blacks in America who doubted the value of busing, but only black nationalists and separatists, it seemed, were allowed to say so in public. The rest of us, we were told, didn't speak for blacks, thus making us irrelevant. Those who dared to speak out anyway were written off as sellouts or Uncle Toms. Why, I asked myself, did the advocates of busing answer all objections to their views not with reasoned arguments but contemptuous ad hominem attacks—and what made the media so willing to go along with them? I couldn't see what made these people so sure of themselves, yet whenever I shared my doubts with a supporter of busing and pressed him to explain why I was wrong, I was fobbed off with the non-argument that no black person had a right to think the way I did. The popular political answers of the day, I saw, had hardened into dogma, making anyone who questioned them a heretic. Having turned my back on religion, I saw no reason to accept mere political opinions as gospel truth. Years later these same dogmatists would walk away from the wreckage of their failed policies, like children tossing aside a broken toy. But the victims they left behind were real people—my people.

Loneliness breeds doubt, and the absence of dissenting black voices from the debate over busing was so complete that I found it hard at times not to doubt my own convictions. Might I have failed to think things through completely? Had I somehow lost touch with my roots? I needed to have my views confirmed by someone I could take seriously. When that confirmation came, though, it was from the least likely of places. Dick Wieler and I had always been frank with each other about racial matters: I told him my innermost thoughts, and he knew how isolated I felt. Shortly after I moved to the civil division, he took a job at the Department of Revenue, where he worked for Janet Ashcroft, John's wife and the general counsel. I often walked over there to eat lunch with him, either in his small office or in the cafeteria. After one of our lunchtime conversations about race, Dick called me at my office. "Clarence, you're *not* alone," he said gleefully. "Take a look at the book review in this morning's *Wall Street Journal*. It's about another black guy who thinks like you." He offered to loan me his copy the next time I came to lunch, but I couldn't wait that long, so I went right back to his office and picked it up.

I turned to the article, which was by Michael Novak, a man whose name I didn't know. He was writing about a book by Thomas Sowell called *Race and Economics*. His very first words took my breath away: "Honesty on questions of race is rare in the United States. So many and unrecognized have been the injustices committed against blacks that no one wishes to be unkind, or subject himself to intimidating charges. Hence even simple truths are commonly evaded." It was as though he were talking directly to me. At the end he quoted from the last paragraph of Sowell's book:

> Perhaps the greatest dilemma in the attempts to raise ethnic minority income is that those methods which have historically proved successful—self-reliance, work skills, education, business experience—are all slow developing, while those methods which are more direct and immediate—job quotas, charity, subsidies, preferential treatment—tend to undermine self-reliance and pride of achievement in the long run. If the history of American ethnic groups shows anything, it is how large a role has been played by attitudes—and particularly attitudes of self-reliance.

I felt like a thirsty man gulping down a glass of cool water. Here was a black man who was saying what I thought—and not behind closed doors, either, but in the pages of a book that had just been reviewed in a national newspaper. Never before had I seen my views stated with such crisp, unapologetic clarity: the problems faced by blacks in America would take quite some time to solve, and the responsibility for solving them would fall largely on black people themselves. It was far more common in the seventies to argue that whites, having caused our problems, should be responsible for solving them instantly, but while that approach was good for building political coalitions and soothing guilty white consciences, it hadn't done much to improve the daily lives of blacks. Sowell's perspective, by contrast, seemed old-fashioned, outdated, even mundane—but realistic. It reminded me of the mantra of the Black Muslims I had met in college: *Do for self, brother.* Now I began to see more clearly why they had impressed me. Though their religion was starkly different from the Catholicism of my youth, their unswerving belief in self-reliance wasn't so far removed from Daddy's way of thinking, and you didn't have to go along with their racial separatism—I didn't, then or later—to know that blacks could never hope to improve their lives until they took responsibility for them.

As I read Novak's review, I shamefacedly recalled my one previous encounter with Thomas Sowell's writing. While I was at Yale, someone gave me a copy of *Black Education: Myths and Tragedies,* in which he wrote with equal candor about the effects of preferential policies on black college students. At the time I was still trying to come to terms with my own situation at Yale, and it was too soon for me to profit from such straight talk. Instead of giving the book a fair hearing, I skimmed it angrily and threw it in the trash, furious that any black man could think like that. Now I saw the tragedy of my own youthful dogmatism—and I also began to see why so many blacks were so uncomfortable with those who broke with the conventional wisdom on race.

I spent a whole week trying to find a copy of *Race and Economics.* Finally I called a friend who worked at a university bookstore. He tracked the book down for me and I ordered six copies, knowing that I would want to have extra copies to lend or give away. I read it hungrily, underlining key passages and reflecting on their implications. The more I read, the more terrified I grew of the dire, perhaps irreparable consequences of the unthinking adherence to such fashionable policies as school busing. But I didn't see *Race and Economics* as a political statement, nor was it meant to be one: it simply tried to tell the truth about a subject that too many people were unable or unwilling to discuss honestly. Reading it didn't turn me into a conservative, much less a Republican. All I cared about was finding answers, no matter who had them. When, later on, I began to associate with conservatives, it was because their ideas were closer to mine than liberals' ideas, not because I saw myself as one of them. I'd already noticed that it was liberals, not conservatives, who were more likely to condescend to blacks, but I assumed, like the good radical I once was, that liberals and conservatives were simply two different breeds of snake, one stealthy, the other openly hostile. Yet here was a black man who talked hard common sense about race—the same sense I was groping toward—and was being praised for it in America's most staunchly Republican newspaper. All at once the political spectrum looked more complicated than I had previously suspected.

I HAD EVEN more to do in the civil division than in the criminal division. In addition to writing briefs and arguing cases in the appellate courts, I drafted a wide range of pleadings and other court documents, and had bench trials in a number of tax cases. I also wrote official legal opinions for the attorney general and advised the state agencies with which I worked. More often than not, I arrived at my desk by four-thirty, going straight to the basement to turn on the coffeepot. (Sometimes I woke up the janitor, a tipsy fellow whose name, oddly enough, was Clarence.) I used the quiet mornings to dictate briefs or pleadings, usually finishing most of them before the rest of the staff showed up at eight. More than once I talked for hours before discovering that the Dictaphone hadn't been working. I had no choice but to retrieve my notes and start over.

It heartened me that others respected my work. One judge, Robert G. Dowd Sr., of the court of appeals in St. Louis, was so impressed that he sent Attorney General Danforth a letter praising my performance. "You can be proud of the excellent job he does in briefing and in argument," Judge Dowd wrote. "His brief is scholarly, his argument is to the point, and he always demonstrates a thorough knowledge of the cases cited." I read that letter over and over, soaking it in and savoring each word; I put it in the middle drawer of my desk, taking it out whenever my spirits needed a boost. In the years to come I would carry it with me from job to job, like an heirloom, rereading it on the dark days when I was racked by self-doubt. I'm sure I made more of it than Judge Dowd had intended, but it meant more to me than any compliment I had ever received, and there would be more than one time down the road when I would need it badly.

Life was good in Jefferson City. Attorney General Danforth made no secret of being pleased with me. Kathy had graduated from college with honors, and Jamal was now a healthy four-year-old. My salary continued to go up bit by bit, though never quite enough to keep me from borrowing pocket money from friends and washing my own shirts every night. I might have stayed on anyway, at least for a little while longer, but when the attorney general decided to run for the Senate, I decided to try something new. This time I had plenty of places to look: the dean of the law school at the University of Missouri in Columbia had invited me to consider joining his faculty, and several attorneys from large law firms in Kansas City and St. Louis who'd seen me in court had asked me to keep their firms in mind if I ever decided to move on. I was pleased by the attention—it was reassuring to find myself in demand at last—but I was more interested in practicing law than in teaching it, and I thought I might feel more at home in the law department of a large corporation, where I'd be able to mix law and business, than in a big-city legal factory.

Once again I spoke to the attorney general, and he put me in touch with George Capps, a businessman friend from St. Louis, who suggested in turn that I approach Monsanto, a multinational chemical company based in St. Louis. I interviewed with Ned Putzell, Monsanto's general counsel, whom I found likable and engaging, and I also spoke to Larry Thompson, the only black member of the company's law department. I'd already met Larry when I was studying for the Missouri bar exam—he was the University of Michigan Law School graduate who told me that he'd left his race off his application—and I liked the idea of renewing our acquaintance. I persuaded myself that Monsanto was the place for me, and when the company offered me a job early in 1977, I accepted. I hated to say good-bye to my friends, but I was sure I'd made the right choice.

Kathy and I rented a small, inexpensive town house apartment in St. Louis that was close to Monsanto's corporate headquarters and to McDonnell Douglas, the aerospace manufacturer where she had landed a job as a systems analyst. We found day care for Jamal, set up housekeeping in our new home, and started getting used to the unfamiliar sensation of having enough money to pay our bills. Then I got a call from my brother, Myers, one morning as I was leaving for the office. "I'm on my way to your house to live with you," he said. "I need directions." He explained that he hadn't been able to get a job in Savannah after graduating from college and had decided to come live with me in St. Louis until he could find work. I was surprised to hear from him—we hadn't seen much of each other since the summer we'd spent living together in Savannah while I worked for Bobby Hill—but he was my brother and needed help, so I told him how to find our apartment and he showed up that night. Myers said that he wanted to work in the hotel business but was willing to take any honest job he could find, and promised to move out as soon as he could afford a place of his own. I never doubted him for a moment. Like me, Myers knew there was no such thing as a dead-end job, and he proved it by finding a job the very next day. I assured him that Kathy and I would pay his living expenses as long as he stayed with us so that he could save all the money he could, and within a couple of months he'd rented an apartment and brought his wife, Dora, to St. Louis.

What followed was one of the happiest times of my life, for it was in St. Louis that I got to know Myers as an adult and a friend. It helped that our schedules were similar, which allowed us to spend a lot of time together. We spent much of it arguing good-naturedly about politics. Myers was still a Democrat, but I now thought of myself as an independent, having grown disenchanted with the pandering and paternalism of the Democratic Party. After one particularly intense argument, I warned him that he'd change his tune as soon as he started paying the taxes to finance his party's pipe dreams. (My prediction came true.) Dora was my kind of person, too, a farm girl from Georgia who had done well for herself in spite of her parents' modest circumstances. An honors graduate in accounting from Savannah State College, she was bright and friendly, and Kathy and I enjoyed her company as much as Myers's. Rarely did more than a few days pass without our seeing them—if they weren't at our place, we were at theirs—and within a year of their arrival in St. Louis, they'd found good jobs and were moving up in the world.

The closer I grew to Myers, the more I longed to know Daddy better. It never happened. I kept on visiting Savannah—at first regularly, then less so—and we became less hostile toward each other, but I found it impossible to relax in his presence. Things were different with Jamal, though. Kathy and I took him to the farm as soon as he could walk so that he could spend half the summer with Daddy and Aunt Tina, and they showered him with love and affection. As far as Daddy was concerned, Jamal could do no wrong. If he wanted to ride on the truck, Daddy took him wherever he wanted to go; if he wanted ice cream, Daddy would drive him to the store; if he wanted to watch cartoons on Saturday morning, the two of them sat down in front of the TV and watched them together. Never had I known my early-rising grandfather to go to bed later than eight o'clock, but now he was happy to sit up with Jamal until ten or eleven.

I knew Daddy had changed beyond my imagining when I drove out to the farm to pick up Jamal and take him back to St. Louis. I went into the kitchen and saw five or six open boxes of cereal, all of them pre-sweetened brands. I couldn't believe my eyes. Daddy had never permitted Myers and me to eat anything other than Kellogg's Corn Flakes, and the idea of opening more than one box at a time was unthinkable. I asked Daddy why he'd let Jamal open so many boxes. He said that Jamal had wanted the prizes stashed in each of the boxes, so he'd opened all of them. "Tell me something, Daddy," I said. "You never make Jamal do anything he doesn't want to do. You let him do whatever he wants. You do whatever he asks you to do. But you *never* treated Myers and me that way. Why not?" Without hesitating, Daddy replied, "Jamal is not my responsibility." And it really was as simple as that. Daddy had to raise us, but he only had to enjoy Jamal, so he kissed and hugged him, slept next to him, bathed him gently, read to him as best he could, and showed him boundless love. He treated his other great-grandchildren the same way. It was a side of him I had never seen.

It made me envious to watch Daddy playing with Jamal. I couldn't help but feel that he'd cheated me out of something important—but I also wondered how hard it had been for him to hide his affection from us. How often had he looked in on my brother and me as we slept, gazing at us with the same sweetness I saw each time he looked at Jamal? How often had he longed to hold us, hug us, grant our every wish, but held himself back for fear of letting us see his vulnerability, believing as he did that real love demanded not affection but discipline? Aunt Tina never doubted that Daddy loved Myers and me deeply; she told me so time and again, always begging me to try and be more understanding of him. How many times, I now wondered, had she begged him to treat us with the natural warmth he thought it his duty to hold back?

On occasion Daddy would apologetically assure me that he had done his best to raise us right. It was as though he were asking for my approval. "You and Myers never caused me any trouble," he said, contrasting our behavior with that of the kids he knew who had been arrested, had illegitimate children, or used drugs. A time came when he was no longer harsh at all—but my memories of him were. Outwardly I treated him with respect; on the inside I still seethed over old grudges, constantly throwing up barriers to his tentative overtures. I couldn't forget the morning he threw me out of the house, or his refusal to come to my graduation from Yale Law School. His was the one voice of approval I needed to hear at that crucial moment, yet he remained silent, and his hardness had hardened my own heart. Eventually the chasm that separated us became too wide to cross. It is my fault, not his, that I never tried to bridge it. Only in the very last months of Daddy's life did we share a solitary embrace, and by then it was too little, too late. Not a day passes that I don't wish I had thrown open my arms sooner to that good man. Not until he was gone did I know how wrong I'd been to turn away from his love.

MY JOB IN the attorney general's office had started with a bang, but my job in Monsanto's law department began with a whimper. I was assigned to the law group of the Monsanto Intermediate Chemicals Company, and I soon saw that there wasn't nearly enough work to go around. Not only did most of the plum assignments go to outside lawyers, leaving the in-house attorneys to fight over the scraps, but the work we did do was closely supervised. The freedom I had known in the attorney general's office, I discovered, made it all but impossible for me to fit smoothly into the culture of a large corporate-law department—I can't even begin to imagine what it would have been like had I gone to work for a big-city firm—and I grew increasingly sullen and discontented as a result. The one good thing about Monsanto was the money, but by then I knew there were more important things in life.

My disillusion deepened when I noticed that Monsanto employed a number of talented blacks who should have been moving up the corporate ladder far more quickly. I went to the black manager in charge of affirmative-action compliance to complain about his complacent attitude toward these gifted young managers. He informed me that the EEO/Affirmative Action report that he submitted each year to the Department of Labor showed that Monsanto was in full compliance with the law. (I later noticed that he was careful to keep a copy of the latest report handy at all times. It was exhibit A in his defense of the company.) I replied that numbers alone didn't tell the whole story, but he wouldn't budge an inch. We met more than once, always with the same result. In time I figured out that he saw Monsanto's black managers as nothing more than fungible percentage points on a government form: the only thing he seemed to care about was meeting his quotas for the year. I still had a lot to learn about affirmative action.

It wasn't until I was assigned to work on Monsanto's waste-disposal contracts that I took any real interest in my new job. More than a few of my colleagues were amused by the passion with which I embraced the cause of waste disposal, but I felt sure that the company had a hugely serious problem on its hands. I grew ever more concerned as I came to better understand the dangers of some of the highly toxic substances created as by-products of our operations, especially after I started reading the animal studies that were being performed for Monsanto in order to establish the safety of its new products. Given the dangerous nature of the chemicals we manufactured, it stood to reason that we had an ethical obligation to minimize or eliminate any harm resulting from the products we made and the processes by which we manufactured them. Such efforts, I knew, were bound to occupy much of the company's time and resources in the future. "Garbage is the future," I said whenever I was dismissed as an alarmist.

As I learned more about the physiological effects of human exposure to toxic waste, I uncovered the answer to a question that had been puzzling me for years. Back when I was going to Saint John Vianney, Mr. Cohen, our next-door neighbor in Savannah, had become mysteriously paralyzed on one side of his body. Daddy said that he'd suffered a "pin stroke," a term that was new to me (I've never heard anyone else use it since). The paralysis was permanent, and this giant of a man spent the rest of his life sitting on the front porch of his house, swatting flies with his one good arm. Before he fell ill, Mr. Cohen had worked in a plant where creosote, a wood preservative, was baked into utility poles to keep them from rotting. One of the studies I read showed that human exposure to creosote could cause the same kind of neurological injury that left Mr. Cohen paralyzed. I remembered Daddy complaining about the foul-smelling "tar" with which Mr. Cohen worked, but it had never occurred to me that it might have made him sick.

How many other hardworking people, I wondered, had been robbed of their livelihoods because of the toxic chemicals manufactured by companies like Monsanto? The more I considered their plight, the more I longed to go back to Savannah and help them. Even then, though, I cared about people, not theories. I had no wish to spin individual cases into some grandiose, ideologically driven legal theory. I no longer believed in utopian solutions, or the cynical politicians who used them to sucker voters, claiming to care about the poor while actually exploiting them. Not only was I sure that such solutions were doomed to failure, but I also feared that once they failed, the resulting disillusionment would make matters even worse. Yet it was taken for granted in the seventies that the purveyors of these elaborate nostrums were doing the right thing, and anyone who dared to challenge their effectiveness was hooted down. That prospect intimidated me, especially when it came to racial matters. If a distinguished scholar like Thomas Sowell could be dismissed as not really black for daring to challenge the liberal orthodoxies on race, what would happen to me if I dared to agree with him? I knew that until I was ready to tell the truth as I saw it, I was no better than a politician—but I didn't know whether I would ever be brave enough to break ranks and speak my mind.

What I did know was that I'd made a mistake by going to work for Monsanto. Things improved somewhat when I was transferred to the agricultural-products group, where Wayne Withers, my new supervisor, gave me the freedom I needed to do my job on my own. Even then, though, there wasn't enough work to keep me busy, and I ended up spending my afternoons reading *Forbes* and *The Economist* and working my way through tall stacks of books about business and governmental policy. I learned a lot from my afternoon reading, but my job rarely gave me the opportunity to put what I'd learned to good use.

Just as worrisome was the fact that Kathy and I were sinking into a financial rut as ominous as the professional rut in which I was stuck. My rickety old Volvo finally gave up the ghost one snowy morning, leaving me stranded in a snowdrift, partially blocking a heavily traveled street. I slid to work in the snow and sludge, thinking with each soggy step of a saying of the old folks in Savannah: "The harder I work, the behinder I git." I got even farther behind when we yielded to temptation in 1979 and bought a house on a corner lot. The purchase saddled us with a monthly mortgage payment of $550, almost twice what our rent had been. Though our income had gone up considerably, the combined weight of taxes, living expenses, and accumulated indebtedness kept us in the hole.

The harder *I* worked, the emptier I felt. I had come to Monsanto looking for success, but found that personal drive and motivation weren't always rewarded in so large and cumbersome an organization. In time I saw that my personal definition of success was as mistaken as the decision I'd made to go to work there. I had manufactured artificial goals as a means of motivating myself, using my longing for money, cars, and other material possessions to create a false sense of purpose. They had worked on me like spoonfuls of sugar—a jolt of energy that soon faded, leaving behind the pangs of a deeper hunger. I had cut myself off from the transcendent hope of religion, and now a vast and frightening expanse of uncertainty lay before me. "The man ain't goin' let you do nothin'," my Savannah friends had warned me long ago. For years I had done all I could to ignore those negative, spirit-sapping words, but they echoed in my mind like the voice of fate. I had done so much to gain so little. Why was I even trying?

My sense of hopelessness was intensified by the discontent I felt with my marriage. I'd always had doubts about the wisdom of marrying so young, but I forced myself to ignore them, hoping that things would somehow work themselves out. They never did. Kathy was a good wife and mother, but I had nothing to offer her beyond the mere desire to be an equally good husband. I was powerless to command my feelings, yet I knew that to admit them would hurt two blameless people who were closer to me than anyone else in the world. It was an impossible situation, and like so many people who find themselves in such situations, I sought comfort in the bottle. I can't think of a single day during my years at Monsanto on which I didn't have at least one beer, and usually many more. On weekends and holidays, I would spend the whole day and most of the evening downing one beer after another: I drank when I watched TV, when I barbecued, when I washed the car, whenever friends came over. I wasn't drinking alone—not yet—but I was definitely drinking too much.

I yearned for guidance from someone with whom I could identify, but I didn't know where to get it. I wasn't close enough to Daddy to talk to him honestly about my problems, and Bobby Hill had turned out to be a hopelessly flawed role model. The only figure from my past for whom I still felt undiminished respect was Father Brooks, now the president of Holy Cross College. In 1978 he had reached out to me in a wholly unexpected way, calling me up to ask if I would consider becoming a member of the school's board of trustees. I was honored and flattered by the request, and naturally I agreed. I had long admired Father Brooks, but I came to revere him during the many years I spent on the board, and to appreciate more fully the differences between Holy Cross and Yale Law School. At Yale I felt I was part of some grand plan that was more about the school than me. Not so at Holy Cross. Father Brooks wasn't trying to prove a point or make a statement by helping us. He knew we were in a tough situation, and he wanted to make it work. That was the kind of help I needed now—but where was I to find it?

As my thirty-first birthday approached, I took a day off and spent it in the law library of the St. Louis County Bar Association in Clayton. The room was bright, quiet, and almost empty, and I left only once, to take a walk and get a bite to eat. All I brought with me was a pen and pad, for I had gone there to write down my thoughts about what I wanted out of life. Popularity? That had never meant much to me. As for prestige, I'd gone to Yale Law School partly because of its reputation, and now I knew all too well what that was worth. Money was different—I had debts to pay—but working at Monsanto had shown me that it alone would never be enough to satisfy me, and I promised myself that I would never again take a job merely for the money. What I cared about more than anything else, I decided, was the condition of blacks in Savannah and across America. The only way I could hope to find personal fulfillment was to spend the rest of my life trying to make their lives better, and to do so in a manner that was consistent with the way Daddy had raised me. As a young radical, I had found it easy to cloak my belief in the necessity of black self-reliance in the similar-sounding views of Malcolm X and the Black Muslims. It wouldn't be so easy now.

To unhesitatingly proclaim the rightness of Daddy's way of life would be to court the same kind of ridicule to which Thomas Sowell was being subjected. Once I, too, had fallen into the trap of condemning as Uncle Toms those black people who thought the way he did. I knew the same ridicule would greet me if I took his side—but I also believed he was right. Though I feared the consequences of saying so publicly, I knew that someday I would have to confront that fear.

Driving home from work the next day, I stopped at the intersection and took a long look at my splendid new home, a handsome ranch house with a stone facade and an expansive front lawn that gradually sloped down from the house on the two sides facing the streets. Never before had I lived in such a place, or thought I ever would. Wayne Withers had just told me that I was going to be promoted and that I was now eligible to participate in the company's bonus program. The hard times would soon be over. I thought of the run-down shanty in which I had been born, and the house in Liberty County that I had helped to build. Pinpoint seemed far away, but it had never been closer. I sat in the car for what seemed like minutes, though it was probably only a few seconds. I could feel the golden handcuffs of a comfortable but unfulfilling life snapping shut on my wrists. I had to quit now—or I never would.

A few days later, I received a phone call from Alex Netchvolodoff. We had worked together in Jefferson City, but he had gone to Washington in 1976 to work for Jack Danforth, who was now the junior senator from Missouri, and we hadn't spoken since I moved to St. Louis. After a few preliminary pleasantries, he got straight to the point: Senator Danforth wanted me to come join his staff in Washington. I said I was interested, so long as I wouldn't have to work on civil-rights issues or matters involving race. Though I cared deeply about these issues, I knew I wasn't yet ready to expose myself to the bruising criticism that would follow once my views became known. Netch told me that he couldn't match Monsanto's overall compensation package or pay for my move to Washington, but he could offer me the same base salary that Monsanto was paying me. I understood what that meant: not only would Kathy have to give up her job at McDonnell Douglas, but the cost of living in Washington was much higher than in St. Louis. If we moved there, the penny-pinching would start all over again. I didn't care. Monsanto had everything to offer me, and nothing I wanted. This was a chance to go east, closer to home, the place where I longed to be. I couldn't say no.

6

A QUESTION OF WILL

My departure from Monsanto was much easier—and more festive—than I'd expected. As soon as Wayne announced that I was leaving, Dick Duesenberg, the general counsel, asked me to drop by his office. Underlings like me rarely ventured into the D building, where the top executives worked, and it was the first time I'd been there since my interview two and a half years earlier with Dick's predecessor. I assumed that the visit was a mere formality and expected him to show no great interest in my decision. Instead he was ebullient, assuring me that I was starting out on an exciting journey that would lead me to unknown heights. On paper his words might have read like platitudes, but I knew from the look on his face that he was sincere, and I never forgot them. Dick didn't forget me, either: we saw each other many times in later years, and he always made it clear that he regarded me as a friend. Wayne arranged a going-away party for me at a nearby restaurant, and both of us were surprised by the number of people who showed up. Not only was the room packed, but everyone present was visibly sorry to see me go. I was less surprised that so many people found it hard to understand why I wanted to leave a secure, well-paying job to work in Washington for less money and fewer prospects. I couldn't blame them for wondering. I knew I had to leave Monsanto, but I also knew I was taking a big chance.

In August 1979, I loaded our furniture into a U-Haul truck and drove halfway across the country to our latest residence, a house in Maryland that I'd rented sight unseen. It was one of the small, inexpensive tract homes built in huge numbers in the fifties, and its best days were long gone by the time we moved in. The yard was small, the shrubbery overgrown, while the shaded walkway to the house was slippery with moss. The carpets were damp, stained, and worn. The filthy venetian blinds on the windows were at least a foot too long and the slats were pocked with rust. The refrigerator was full of mold, the bathroom was poorly cleaned, and the basement was damp and barely habitable. It wasn't much of a home, but we were stuck with it. We were also stuck with a pile of bills that I couldn't pay. Rent, the cost of moving, the mortgage payments on our house in Missouri, Jamal's private school tuition: it all added up to an appallingly heavy load, and Kathy had yet to find a job in Washington, meaning that for the moment we would have to make ends meet on my salary alone. Several months went by before we managed to find a buyer for our house in St. Louis, and I called our real estate agent every few days, hoping against hope for good news. I grew sick with worry, but no matter how bad things got, I had no intention of letting Jamal go to a public school, so Kathy and I trimmed our household budget to the bone and held our breaths.

The only thing that kept me going during those first few months was the sheer excitement of living in Washington. I loved riding the bus to work every day—I got goose bumps the first time I passed by the White House—and it took a long time for me to get used to the fact that I was working on Capitol Hill. Sometimes I would go outside and spend a few minutes gazing at the Capitol dome, just like Jimmy Stewart in *Mr. Smith Goes to Washington*, and I also liked to stroll over to the Library of Congress and do research in the beautiful reading rooms. Once I stopped to peek into the majestic Great Hall of the Supreme Court Building. I was almost afraid to walk in, but I finally passed through the front door and spent a few breathless moments gaping at the marble columns and the busts of the chief justices. It was hard to believe that I was standing in a place no member of my family had ever seen, and that I had never hoped to see. Young people today use the word "awesome" casually, but I don't know any other way to describe the impression it made on me: I was filled with reverence and wonder. My first trip to Washington had been to protest the Vietnam War, but now I was part of the decision-making apparatus of the federal government—the tiniest possible part, to be sure, but a part all the same. I was no longer on the outside looking in. You can't live in Washington for very long without becoming cynical about politicians and their motives, but I've never doubted the greatness of a country in which a person like me could travel all the way from Pinpoint to Capitol Hill.

I didn't have much time for sightseeing. Senator Danforth had given me responsibility for energy, the environment, and public works, and my plate was as full in Washington as it had been in Jefferson City. President Carter had just declared war on the energy crisis, and the capital was full of talk about what the government could do to solve it. The days when I had to scrounge for work every morning and read magazines all afternoon were gone for good, though I went out of my way to find time to learn more about the condition of blacks in America, the issue I cared about most. I had learned at last to think calmly and analytically about racial issues, and I put no stock in party labels. All I cared about was solving problems. I was happy to listen to anybody who was capable of talking sense instead of spouting tired slogans.

One of the first people in Washington who talked sense to me about race was Jay Parker, the editor of a new magazine called the *Lincoln Review*. Chris Brewster, an old friend from the Missouri Attorney General's office who had come to Washington to work for Senator Danforth, told me that Parker sounded like a man I might find interesting, so I called him. I expected him to give me the runaround, since I was a lowly legislative assistant, but we talked for an hour on the phone, then made a date for lunch. Gil Hardy, who had moved to Washington after a two-year clerkship for a judge in the Virgin Islands, joined us for lunch at Duke Ziebert's old steak house on New Year's Eve, and the three of us chatted in the nearly empty dining room for several hours. Jay was friendly, energetic, unflappable, and unapologetically conservative. I'd never known a black person who called himself a conservative, and it surprised me that we rarely disagreed about anything of substance.

Most of the blacks I met in the Senate, even the ones who worked for Republicans, were anything but conservative, though that didn't stop me from getting to know them. Among the black Senate staffers whom I got to know well were Ralph Everett, who worked for Fritz Hollings, a Democrat from South Carolina; Ron Langston, who worked with Roger Jepsen, a Republican from Iowa; and Tony Welters, who worked for Jacob Javits, a Republican from New York. In addition there were also a small number of black staffers working for Republican House members. Someone suggested that we establish a group for black Republican staffers, which led to the founding of the Black Republican Congressional Staff Association. I was neither a joiner nor a Republican—yet—but I made an exception for BRCSA, since it helped to relieve the isolation that each of us felt working in our separate offices. It wasn't exactly fashionable to be a black person working for a Republican, and it was comforting to meet others in the same boat.

My new friends and I also made an effort to develop relationships with black staffers on the Democratic side of the aisle, many of whom viewed us with a mixture of curiosity and disapproval. Since nearly all of us were political moderates or outright liberals, I couldn't see why they should be so disapproving, but they were, not because of our political views but simply because we were willing to work for Republicans. This sentiment ran deep among blacks. When Coretta Scott King visited Senator Danforth to ask for his assistance with a fund-raiser for the Martin Luther King Foundation, Alex Netchvolodoff asked me to accompany her to the Capitol to meet with the senator. After the visit was over, I walked Mrs. King to her waiting car, mentioning on the way that the senator was a good man who cared deeply about the plight of blacks in America. (He had a long, close relationship with Morehouse College, a historically black college, and had been a member of its board of trustees.) I lamented that so few blacks had voted for him. "Well, he *is* a Republican," Mrs. King replied. I knew at once that I'd just learned something important about the way most blacks felt toward the Republican Party.

It didn't surprise me, for I, too, had reflexively disliked most Republicans (John Bolton had been an exception) prior to going to work for Senator Danforth in Jefferson City. Still, I found it hard to accept. "Black is a state of mind," one black Democratic staffer told me, by which I assumed he meant being a liberal Democrat. That kind of all-us-black-folks-think-alike nonsense wasn't part of my upbringing, and I saw it as nothing more than another way to herd blacks into a political camp. But some of the black staffers I met on Capitol Hill were so sensitive to criticism that they found it necessary in conversation to elaborately disavow the positions of the Republican congressmen or senators for whom they worked, and I liked that even less. If I hadn't felt at ease with Senator Danforth's political views, I would have found myself another job.

I saw a memorable demonstration of this kind of hypocrisy when Thomas Sowell met with a group of black Capitol Hill staffers in 1980. By that time I'd read most of his books and many of his articles, and I wasn't surprised to hear him express in no uncertain terms his opposition to school busing, racial quotas, and increasing the minimum wage, all of which he believed to be bad for blacks. His bluntness ruffled more than a few feathers; in the eighties these were all hot-button issues for black politicians and community leaders. What bothered me was that some of the questions from the Republican staffers were both hostile and embarrassingly uninformed. That didn't fluster Sowell. (As I was to learn, nothing flusters him.) He defended his positions with cool, clearheaded logic, and though his questioners grew emotional, none of them was able to refute his arguments.

Hearing Sowell, and speaking to him privately after the meeting, was a landmark event for me. So was my first encounter with Walter Williams, which took place not long afterward at the Heritage Foundation, a fairly new conservative think tank. Dr. Williams was an economics professor at Temple University who had grown up in the tough part of Philadelphia, driven a taxi, served in the Korean War, and earned a doctorate in economics from UCLA. Street smart and quick witted, he had written extensively on the negative effects of government regulations on blacks, a subject in which I was closely interested. Among other things, his research had proved that New York City was unintentionally keeping blacks out of the taxi business by placing a strict upper limit on the number of medallions in circulation (a medallion is required to operate a taxi), thus raising the entry cost of driving a cab and preventing poor blacks from entering the business. Very few black scholars were using that kind of research-driven thinking to study the everyday problems of blacks, and Dr. Williams's findings were as exciting to me as they were upsetting to those who still believed that government regulation was the only way to improve the lot of black people.

Walter Williams, Thomas Sowell, and Jay Parker were all smart, courageous, independent-minded men who came from modest backgrounds. Politics meant nothing to them. All they cared about was truthfully describing urgent social problems, then finding ways to solve them. Unhampered by partisan allegiances, they could speak their minds with honesty and clarity. They were my kind of black men. I got to know Jay best, however, because he lived in northern Virginia and kept an office in Washington, making it easier for us to meet regularly. We became fast friends, and before long I started confiding in him about my personal difficulties. He treated me like family. I liked his straightforward, positive attitude, and I also appreciated the fact that he didn't look to politicians to fix his life, not even the ones he admired most. (I'll never forget the time when he reminded me that freedom came from God, not Ronald Reagan.) For Jay politics was a part of life, not a way of life. It was an attitude I sought to emulate.

If only my private life had been half so satisfying as my professional life, I would have been a truly happy man. But moving to Washington had done nothing to ease the dissatisfaction I felt with my marriage. I hated myself for my inability to be the loving, devoted husband Kathy deserved, yet I couldn't seem to do anything about it. I grew despondent and resigned, so much so that I actually stopped worrying about how to get out of debt and achieve financial security. Such things no longer seemed to matter. Instead I drank more heavily than ever before, and though I was careful not to let my drinking interfere with my work, I knew I was on the road to trouble.

I didn't realize how far I'd traveled down that road until the summer of 1980, when Butch Faddis, one of my Missouri friends, called me up and suggested that the two of us run the New York City Marathon together. I was overweight and pitifully out of shape, but I decided to give it a try. I started with a short run around the neighborhood. After thirteen minutes I was so close to collapsing that a neighbor asked me if I needed help. Bending over, gasping for air, and clutching the stitch in my side, I somehow managed to tell her that I was fine. I wasn't, of course, not in any sense of the word, but at least I didn't quit. Instead I started thinking of my training as a metaphor for my life. While much of its disorder was beyond my control—I couldn't will myself to be a loving husband, push a button and pay all my debts, or solve all the problems of blacks in America—running in the marathon was a finite, achievable goal that was within my power to achieve, so long as I worked toward it steadily and methodically.

I took as my motto a saying of Bobby Knight, then Indiana University's men's basketball coach: "Everybody has a will to win. What's far more important is having the will to *prepare* to win." The question, I saw, was whether I had the will to prepare for the marathon—and ultimately, for the rest of my life. I didn't know the answer, but I knew that simply making the attempt would help me. Besides, hadn't Daddy and Aunt Tina done their best to prepare my brother and me for a future they knew they wouldn't live to see? And wasn't that what I was doing with my own son? Running the marathon was simple by comparison. The distance, 26.2 miles, was a constant: I was the variable. I'd had the will to prepare as a student, and as a young attorney in Jefferson City. Soon I would see whether the struggles of the past few years had drained it out of me.

Butch learned that we couldn't run in New York without first establishing a qualifying time in another race, so we agreed to enter the fifth annual Marine Corps Marathon, which is run in Arlington and Washington every fall. I figured I would need to run for one hundred consecutive days in order to make myself ready. I started getting up at four every morning to run, increasing my time by a minute or so each day. Early in my training I saw another runner coming up behind me with a smooth, consistent pace. He closed in and passed me in a flash, breathing easily, striding effortlessly, and greeting me with a friendly wave. I jogged on painfully, panting like a beached whale. I was humiliated, but I didn't quit, and by October I was running ten miles a day. My cheap cotton sweatpants rubbed against the inside of my thighs, causing chafing that wouldn't heal. I carried Vaseline with me to minimize the pain, but there wasn't anything I could do about the soreness in my feet and knees. Yet I kept on running, and one day I saw the jogger who had passed me that summer. This time I passed him, smiling with quiet satisfaction, greeting him with a friendly wave just like the one he'd given me.

That November I joined five thousand other runners at the Iwo Jima Memorial in Virginia. I ran alone—Butch had injured himself and couldn't participate—but I'd lost nearly thirty pounds and felt better than I had in years, and I couldn't wait to get started. Midway through the race, I picked up my pace a bit too much and hit the "wall" I'd heard so much about from my fellow runners. My legs grew stiff and I felt as though I was on the verge of collapsing. *Never quit, never quit*, I whispered over and over again, the same way I had throughout my training, but it didn't help this time. My body begged me to give up with every tortured step, and I barely made it to the next water stop, which was in the parking lot of the Pentagon. A young black Marine was handing out water to the exhausted runners. "God, this is hard," I told him. "That's what you asked for," he replied without a trace of sympathy. I shook off my self-pity, picked up my pace, and crossed the finish line three hours and eleven minutes after I'd started.

I was pleased with my time—the average run time for the Marine Corps Marathon is just over five hours—but even more pleased that I'd had the will to get myself in shape in the first place. As grueling as the experience had been, I found in it a kind of honesty and simplicity that was rare in Washington. Either I prepared each day, or I didn't; either I paid the price and suffered through the pain, or I didn't. Politics was different. It was about perceptions and promises, most of them empty. *Tell me where it hurts and I'll make it better*, the ambitious politician smoothly assures his constituents. Not so the young Marine I met in the Pentagon parking lot. He offered me nothing but a cup of water and a stern reminder that the pain of individual effort is part of the price you pay for achievement; too many of the politicians on Capitol Hill, by contrast, saw it as little more than grist for their electoral mills. Daddy, like the Marine, had known better, and so did I.

IN THE FALL of 1980, I changed my voter registration from Missouri to Maryland—and registered as a Republican. I had decided to vote for Ronald Reagan. It was a giant step for a black man, but I believed it to be a logical one. I saw no good coming from an ever-larger government that meddled, with incompetence if not mendacity, in the lives of its citizens, and I was particularly distressed by the Democratic Party's ceaseless promises to legislate the problems of blacks out of existence. Their misguided efforts had already done great harm to my people, and I felt sure that anything else they did would compound the damage. Reagan, by contrast, was promising to get government off our backs and out of our lives, putting an end to the indiscriminate social engineering of the sixties and seventies. I thought that blacks would be better off if they were left alone instead of being used as guinea pigs for the foolish schemes of dream-killing politicians and their ideological acolytes. How could I not vote for a man who felt the same way?

Reagan won in a landslide, with coattails long enough to sweep a Republican majority into the Senate. The Hill was abuzz with excitement the next day, and my fellow Republicans made no secret of their delight. Conservative groups began to circulate policy papers suggesting possible initiatives for the incoming administration. The best known and most widely read of these documents was the Heritage Foundation's *Mandate for Leadership*, a thousand-page tome that would be carefully studied by Reagan's aides, many of whose policy ideas grew out of recommendations originally made by Heritage fellows. Stuart Butler, one of the policy analysts at Heritage, had already published an essay advocating what he called "enterprise zones." Butler thought that cutting regulation and taxation in inner-city areas would encourage the growth of local businesses, thus improving the quality of neighborhood life. His idea made sense to me, but when I talked it over with Jay Parker, he immediately asked whether I thought inner-city crime rates were too high for businessmen to be willing to relocate to an enterprise zone, no matter how enticing the financial incentives might be. I had no answer.

Shortly after the election, I attended a Senate reception at which Jack Kemp, a well-known Republican congressman from upstate New York, was giving a talk on enterprise zones. During the question-and-answer session that followed his talk, I asked Congressman Kemp the same question Jay had asked me. He fumbled it badly, falling back on a pre-rehearsed sound bite about the plan's advantages. This was my first indication that conservative Republicans, even though they were asking many of the right questions, might not necessarily have all the answers. Yet I remained optimistic about the overall thrust of the Republican Party's approach to urban problems. For the first time in my adult life, Washington was full of serious talk about the possibility of getting government off the backs of the poor. I began to think that something might finally be done to help the people I cared about most.

Not long after that, I received a phone call from Thomas Sowell, who asked me to attend a San Francisco conference on race in America. The purpose of the conference, he explained, was to stimulate new thinking about the issues confronting blacks, and he wanted me to sit on a panel that would discuss education policy. I accepted with pleasure, and in early December I flew to the West Coast—another first for me. The conference was held at the Fairmont Hotel, which seemed to be bursting with energy by the time I got there, though not all of it was positive. Some of the people I met in San Francisco were "8A Republicans," blacks who ran businesses that received preferential treatment under a program started by the Nixon administration. The goal of the program had been to give a temporary boost to black entrepreneurs, but like so many similar undertakings, it had hardened over time into a bureaucracy whose beneficiaries had no intention of giving up their privileges so long as they could hold on to them. These businessmen had no interest in Professor Sowell or his ideas, for they knew perfectly well that he thought the 8A program should be terminated as soon as possible. All they cared about was wangling access to the incoming administration and lapping up the favors it would hand out. I did my best to ignore their minor-league wheeling and dealing, spending my time with those who had come to San Francisco for more idealistic reasons.

Ed Meese, one of President Reagan's top advisers, spoke at lunch on the last day of the conference. A young black reporter from the *Washington Post*, Juan Williams, was seated next to me, and we struck up a conversation. I had no idea that what I said might find its way into the *Post*—I wasn't used to talking to reporters—and I spoke to Williams in a straightforward, unguarded way, explaining that I was opposed to welfare because I had seen its destructive effects up close in Savannah. Most of the older people among whom I had grown up, I told him, felt as I did, sharing Daddy's belief that it would be the "ruination" of blacks, undermining their desire to work and provide for themselves. I added that my own sister was a victim of the system, which had created a sense of entitlement that had trapped her and her children. I went on to say that I opposed busing, preferring to give school vouchers to poor children trapped in dysfunctional schools. We shook hands and parted and I thought no more about it. All I'd done was speak my mind. What could be wrong with that?

I flew back to Washington full of excitement. Then I read the newspaper coverage of the conference, most of which appeared to dismiss our motives as self-serving. One reporter referred to me in passing as a "black conservative." I'd never been mentioned in a newspaper, nor had anyone ever called me a conservative. "I've been called a lot of things in my life," I later told Professor Sowell, "but never that." He laughed. "At least it's better than being called a transvestite," he replied dryly. A bigger surprise was on the way: Janet Brown, Senator Danforth's press secretary, called to tell me that a *Washington Post* photographer was coming by the office to take my picture for a column by Juan Williams that would run in the paper the next day. I almost fainted. I had no idea what the column would say, but I expected the worst.

After a sleepless night, I got up first thing in the morning to buy a copy of the *Post*. I can still remember the date: December 16, 1980. I turned to the opinion section, saw my smiling face, and knew that my life would never be the same. It wasn't that Juan Williams had misquoted or misrepresented me. He presented my opinions accurately and fairly. But I'd gone against the liberal consensus on race, something that blacks weren't supposed to do—and in the *Post*, no less! For the next few days, strangers on the street glared disapprovingly at me as I walked by. Senator Danforth had no problem with the piece—he'd heard me say such things many times before—but I took plenty of heat from some of my fellow black staffers on the Hill, who made it clear that there could be no real debate on these matters. I could only choose between being an outcast and being dishonest. I also received a piece of anonymous hate mail claiming that I wore a "watermelon-eating grin" in the picture the *Post* ran alongside the story.

That was bad enough, but what really hurt was the criticism I received for having mentioned my sister and her children. I didn't blame Juan Williams—it was my fault for not knowing the rules of Washington journalism—but I wouldn't have said anything of the kind had I known he was going to print it. The only reason I'd brought the subject up in the first place was to make clear to him that I knew what I was talking about. What I found inexplicable was that so many of the people who went out of their way to tell me how strongly they disapproved of my views seemed to think that the mere act of pointing out the human damage caused by welfare policies was wrong in and of itself. Would they have felt the same way if I'd said that I was opposed to drunk driving because my sister and her children had been hit by a drunk driver? I doubted it.

Not long after the column appeared, Kathy and Jamal went to Worcester to spend Christmas with the Ambush family. I stayed behind in Washington. Christmas no longer meant anything to me, and I preferred putting in extra time at the office to celebrating a holiday about which I no longer cared. I started drinking as soon as they left. I woke up sick and depressed early the next morning. All I could think about was the angry reaction to the *Post* column. It made no sense to me. Why was it wrong for me to speak my mind? All at once I felt an overwhelming desire to drive down to Savannah and see my family. I didn't understand why—Daddy and I were as distant as ever—but somehow I knew I needed to be with them. I threw my clothes into a suitcase, grabbed a six-pack from the refrigerator, and headed out the door. Freezing rain had fallen during the night and the windshield of the car was thickly covered with ice, but that didn't stop me. I chipped it off and headed south, drinking beer and watching other cars slide off the road and crash into one another.

I had more than enough to keep my mind busy as I made my way to Savannah. How, I asked myself, could I possibly hope to lead an honest life? To be honest about race meant that I would be subjected to punishing personal attacks from all sides; to be honest about the emotional emptiness at the center of my marriage would hurt the two people who loved me most. In order to do what others thought right, I'd have to be dishonest with myself—but to do what I thought right seemed to me to be narcissistic and selfish. Why shouldn't I continue to sacrifice my own feelings for the sake of my family? That was what I wanted to do, but I no longer had the strength to keep on doing it. I barely had enough strength to keep myself going. The more I thought about my situation, the more clearly I understood that I had to leave Kathy in order to survive.

Visiting Savannah did little to take my mind off my troubles. I spent most of my time at the farm, eating and drinking far too much. My family had heard about the *Post* story, but none of them, not surprisingly, had actually read it. When I told them what I'd said, everyone had the same reaction: they couldn't see why anyone had disagreed with me, much less criticized me, since everything I'd said was obviously true. Even my sister agreed, though I apologized to her anyway.

I returned to Washington feeling no better than when I'd left. I was relieved that my family hadn't been hurt by my candor, but I had something far more consequential on my mind, and I knew I didn't dare wait any longer to deal with it: I left my wife and child. It was the worst thing I've done in my life, worse even than going back on my promise to Daddy that I would finish my seminary studies and become a priest. I had broken the most solemn vow a man can make, the one that ends . . . *as long as you both shall live.* I still live with the guilt, and always will.

I moved out a few days later. Gil Hardy let me sleep on the floor of his living room, which became my bedroom for the next few months. I put my clothes in plastic garbage bags, took them to his apartment, and stored them in one of his closets. I was so broke that I often had to borrow small sums of money from friends in order to take the bus to work. Gil was a true friend—I've never had a better one—but he was also a vegetarian who fasted regularly, which meant there was rarely anything to eat in his kitchen. Each night after work I paused at the corner of Sixteenth and K streets, trying to decide whether to take a bus the rest of the way back to Gil's place or grab a bite to eat at the nearby Burger King. If I ate, I walked; if I rode the bus, I went hungry. I spent my days trying to behave as if nothing were wrong, but my friends knew better. Now that the Republicans were in the majority in the Senate, we were given more office space, and Chris Brewster and I had moved into an old converted hotel near the Dirksen Senate Office Building. Sometimes I couldn't muster up enough energy to do anything but sit numbly at my desk. Years later Chris told me that my despair was so obvious, it made him want to cry. I was so unhappy that I started going to church again. St. Joseph's on Capitol Hill was only a short walk from my new office, and I went there each weekday to ask God to give me the wisdom to know what was right and the courage to do it—though I still couldn't bring myself to go to Sunday Mass. I wasn't yet ready to take that leap of faith.

I had always enjoyed chatting with one of the black janitors in the Dirksen Senate Office Building, exchanging greetings and small talk. One day he saw the look of anguish on my face and said something I've never forgotten: "You cannot give what you do not have." His words hit home. My dream had always been to help my people, but how could I tell them not to quit if I gave up myself? How could I convey a message of hope if I was hopeless? How could I face Jamal if I couldn't face myself? I had to pull myself together and try to mend my broken life. It was, like every other obstacle I had overcome, a question of will. Now I would find out how much I really had.

Gil listened to music each morning as he got ready to go to work, and in 1981 one of his favorite songs was George Benson's "The

Greatest Love of All," the theme song of a 1977 movie about Muhammad Ali, whom Gil and I both admired. I'd heard the song many times, but it had never meant more to me than it did now: *The greatest love of all / Is easy to achieve / Learning to love yourself is the greatest love of all.* It was a powerful message for me to hear, since I didn't even like myself. I had good reason not to. My own brother had told me that he couldn't understand how I would do to my son what our father had done to us—and I knew he had a point. Many of my black colleagues on the Hill treated me like a pariah for having broken ranks with them—and I agonized over whether it was right for me to have done so. I could barely bring myself to look in the mirror, much less love myself. "That's what you asked for," the Marine in the Pentagon parking lot had told me. I was getting what I'd asked for, but I'd never imagined that it would hurt so much. Still, I'd done what I thought was right, and I took heart from George Benson: *I decided long ago, never to walk in anyone's shadows / If I fail, if I succeed / At least I'll live as I believe / No matter what they take from me / They can't take away my dignity.* I'd only just begun to take the measure of the price I would pay for going my own way—but at least I would live as I believed.

IN MARCH I received a call from the Office of Presidential Personnel. The request was simple: would I consider becoming the assistant secretary for civil rights in the Department of Education? I said I'd give the offer some thought, but that was mostly just to be polite. While I knew that becoming an assistant secretary at the age of thirty-two would be a smart career move, I expected to say no. Not only did I believe, as did Ronald Reagan, that the Education Department should be abolished, but I also didn't care to work in a civil-rights post. I had no background in that area, and was sure that I'd been singled out solely because I was black, which I found demeaning. Having felt the lash of public criticism, I questioned whether I had the strength—or the courage—to stand in the eye of the howling storm that surrounded civil-rights policies. Besides, I still hoped to move back to Savannah sooner or later, so I couldn't see the point of jumping from Senator Danforth's office to yet another Washington job.

I called up Jay Parker, knowing that he'd tell me exactly what he thought. "Aren't you tired of just *talking* about problems, Clarence?" he asked. "Wouldn't you like to try doing something about them for a change? It's your choice, but if you aren't willing to take the job, then maybe you should stop spouting off about racial issues. Put up or shut up." That made me think twice. I asked Alex Netchvolodoff what he thought I should do, and he told me to take the position. So did Senator Danforth. But Jay's blunt words made the strongest impression on me. It wasn't enough merely to talk about race: I had a moral obligation to see if I could put some of my ideas into action. I agreed to speak to Terrel Bell, the secretary of education, and when he offered me the job, I decided to give it a try.

Secretary Bell asked me to work in the job as a consultant while I waited for the Senate to confirm me, and in May I left the Hill and went to work for the Reagan administration. I was understandably apprehensive about joining an administration that was then being branded as "racist" by so many prominent black leaders; I also had mixed feelings about leaving Senator Danforth's office, for it was a safe haven, a warm, friendly place to work. It had been easy to respect Jack Danforth, an honest man who respected my principles and never asked me to compromise them for the sake of expedience. Unlike most of the elected officials I had seen up close, he wrestled daily with the conflict he saw between being a leader and being a politician. Yet for all his idealism, he was a practical man who knew how to get things done, and I found him wholly admirable. I admired Ronald Reagan, too, but I knew I wouldn't be working directly for him, and I knew little of the political appointees whose job it was to turn his beliefs into policies. They talked a good game, but would they prove to be true to their word?

I couldn't go on sleeping in Gil's living room, so I rented a tiny basement apartment not far from his place, bought a cheap sleeper couch, and moved out. My new home was clean but poorly built. The walls were paper thin and my upstairs neighbors were noisy, and no sooner did they settle down each night than a punctual rat started chewing on the false ceiling over my bed. But I knew I couldn't afford anything better, and my dire straits were brought home to me even more emphatically when I collected my financial records as part of the investigation into my life that was necessary in order to be considered for any government post that, like this one, required Senate confirmation. There was no getting around it: I was on the brink of financial ruin, with no hope of improvement any time soon.

The good news, if you could call it that, was that my new office was a whole lot nicer than my new apartment. It was large and spacious, with a huge desk, a sitting area with a couch, a comfortable conference room, and a private, well-appointed bathroom. It also came equipped with the perfect secretary. Diane Holt, like me, was very young and from Savannah—although in her case it was Savannah, Tennessee. She was smart, mature, efficient, and trustworthy, and she also had plenty of common sense. Diane carefully guarded my door and my time for the better part of the next decade. I needed all the time I could get: I had a lot to learn, and the first thing I had to figure out was who I could trust. Unlike most cabinet agencies, the senior staff at the Office of Civil Rights consisted of career civil servants rather than political appointees, but most of them were very competent and professional, especially Tricia Healy, my special assistant. I worked more closely with Tricia than any other employee, and she became my eyes and ears in the agency.

It didn't surprise me to discover that the senior staff was generally suspicious of anyone from the Reagan administration—President Reagan, after all, had vowed to abolish the Department of Education if elected, though he never followed through on his promise—but I felt no personal hostility from anyone. Most people seemed willing to give me a chance, and I in turn made it clear to them that I wasn't there to impose my own ideological vision on the agency. I was there to solve problems, not advance an agenda, least of all some grand theory about race. I saw my lack of experience as an asset, and for the most part that was what it turned out to be. But I also had little experience in choosing personnel, and one of my first hiring decisions would turn out to be a fateful blunder. Otherwise, like so many other routine personnel decisions, it would not deserve mention.

Shortly after I started work, Gil Hardy called me up and asked me to "help a sister" who was leaving his firm. Her name, he said, was Anita Hill. I asked him if she was a Republican, and when he said no, I warned him that it was all but impossible to shepherd a political appointee through the vetting process unless the person had been a longtime Reagan supporter. But because of my friendship with Gil, I agreed to interview her as soon as I was confirmed. Anita had graduated from Yale Law School the year before, so I started off by asking her why she wanted to leave a prestigious law firm to come to an obscure civil-rights agency. She replied that her options were limited because she couldn't get a recommendation from a partner for whom she'd worked. According to her, the partner had asked her out, and when she declined, he'd started giving her bad work assignments and performance assessments. I asked what she thought of Ronald Reagan. "I detest him," she replied. That meant I couldn't hire her as a political appointee, but Gil had urged me to do whatever I could for her. Not only did I feel I had an obligation to help my fellow blacks, but I remembered how hard it had been for me to land a job after graduating from Yale, and I didn't want to treat her as I'd been treated. I found a way to hire her without going through the nearly impossible hiring process for political appointees. Her work wasn't outstanding, but I found it adequate.

At the time, of course, Anita was the least of my worries: I was preoccupied with my new job, and with my unhappy personal life. Separating from Kathy and Jamal had been more painful than I could bear, and I went back home around the time I started work at the Department of Education. But I knew at once that it was a mistake, and later that summer I moved out again for good. I took my stereo, a damaged table, a chair, and an old mattress, moved into an efficiency apartment close to my office, and buried myself in my work. Secretary Bell was a quiet, even-tempered Mormon who, I soon discovered, was distrusted by many of my fellow political appointees. They questioned the seriousness of his commitment to dismantling the department, but I thought he was doing a good job, and I figured that President Reagan could fire him easily enough if he was truly disloyal. Until then, I decided, I would stand behind him.

When I arrived at the department, Secretary Bell and his staff were in the process of finalizing a number of higher-education desegregation plans. Rather than focusing solely on increasing the percentage of blacks attending the previously all-white colleges and universities—the longtime goal of the influential NAACP Legal Defense and Education Fund—the department was trying to place more emphasis on upgrading historically black colleges. These two efforts, I saw at once, contradicted each other: as more black students started going to white schools, fewer would be available to attend black schools. The leaders of the historically black colleges had privately warned us that the Legal Defense Fund was undermining their attempts to keep these schools afloat. I didn't believe in supporting black colleges that did a poor job of educating their students, but I couldn't see why they should be forced to close their doors in the name of a theory of racial integration that would force blacks to be permanent minorities on predominantly white campuses. To impose mandatory integration policies similar to the ones that had been used in primary and secondary schools seemed to me shortsighted and misguided. My experiences at Holy Cross and Yale Law School had shown me that this approach was no panacea for the problem of black education. Moreover, the historically black colleges and universities had their own traditions, as well as a track record of success. Why, then, wasn't it enough to upgrade them to the same level of quality as the predominantly white institutions, then let black students decide for themselves which kinds of school would suit them best?

One important question often overlooked in the debate was what was actually happening to black students on predominantly white campuses. My class at Holy Cross had contained only six blacks, but none of us failed to graduate on time, and most did very well academically; by the time I joined the board of trustees in 1978, though, very few of the black students who came to Holy Cross graduated in the top half of their classes, and the attrition rate for blacks in predominantly white colleges and universities throughout America was disturbingly high. Almost half failed to graduate on time, if at all. Nor was enough attention being paid to the kinds of courses these students were taking: very few studied math, science, or engineering. To ignore these unpalatable facts was to miss the whole point of higher education. Merely to enroll a black in a predominantly white college means nothing. What matters most is what happens next. An education is meaningless unless it equips students to have a better life.

In one of my early staff meetings, I asked to see any studies that compared the academic performance of black students in integrated primary and secondary schools with black students in segregated or predominantly black schools. None was forthcoming, and when I pursued the matter, a staffer told me that none existed. I asked why it was so widely accepted that black children were better off in integrated schools. He replied that integration had nothing to do with education: the point of busing white and black children to each other's schools was to encourage their parents to move to those neighborhoods. I was aghast, and I had no doubt whatsoever that most blacks would have felt exactly the same way. All the black parents I knew tolerated the disruption of busing solely because they wanted better educational opportunities for their children, not so that they could live next door to whites. Needless to say, I had no problem with desegregating housing, but I couldn't see the point of using black children to bring it about when what they really needed was a decent education. With that, they could decide for themselves where to live.

Alone in my office one evening, I read through the existing reports on the course work and discipline rates for students in integrated schools. They all said the same thing: black students were less likely by far to enroll in the more challenging courses and more likely to have discipline problems. How could they be expected to learn when they weren't even taking the right classes? The data also made it clear that black males were dropping out of high school at an alarming rate, and that those who remained rarely did well academically. (This problem was obscured by the fact that black females performed significantly better than their male counterparts, thus raising the average black performance scores in a deceptive way.) To me the data spelled doom for blacks in America—but I knew that nothing I could say or do about the situation would be heard over the din of dogmatic racial politics. I was overwhelmed by a feeling of hopelessness. Members of my race were caught in a cruel trap not of their own making. My own life seemed to be damaged in a way I didn't know how to repair. It was more than I could take. I sat at my desk and wept.

Part of what overwhelmed me was the knowledge that the disease of blind dogma afflicted both parties. Working in the Reagan administration, you often got the feeling that no black person—especially one who, like me, worked on civil rights policy—could ever be conservative enough for some of the people who had supported the president early on. As far as they were concerned, you weren't a true-blue Reagan man unless you'd backed him at the 1976 convention in Kansas City. That was a test I couldn't ever hope to pass, since I hadn't been politically active in 1976, much less a Republican. What bothered me even more, though, was that some *black* Republicans were expressing doubts about the depth of my commitment. I'd put my reputation on the line by breaking ranks on racial matters and going to work for President Reagan. What more did they want? I appreciated a phone call I got from Sam Cornelius, who worked in the presidential personnel office. Sam, who is black, had been hearing the same whispers, and called to assure me that such comments were nothing more than envious nonsense. "The President put you there, and these Negroes can't do anything about it," he told me. "Don't you worry about them."

I tried not to, but it wasn't easy. I felt as if I were spending half my life sitting in meetings and listening to endless arguments that never went anywhere or did anybody any good. One snowy winter day I walked back to my office from a series of meetings at the main building of the Department of Education. Washington grinds to a halt whenever it snows, and the federal government shut down early that day. I felt bone tired as I trudged through the fast-falling snow, trying in vain to stay warm in my cheap overcoat and leaky shoes. Then I heard from a passerby that a plane had just crashed into the Fourteenth Street Bridge, and that a Metro subway train had derailed. Many people had been killed and many more injured. All at once the long day's meetings seemed unimportant to me. I went home to my tiny apartment and started drinking. That was how I spent too many nights in the early eighties, drinking alone in a dreary efficiency apartment. I called Jamal every night, but he rarely said more than a few words to me. I didn't blame him. I'd let him and Kathy down. What could he possibly have to say to me? I'd started dating again, but the idea of remarrying terrified me, considering the mess I'd made of my first marriage; I was seized with a guilt that I knew would never leave me, and I knew I didn't deserve to be free of it. I hadn't quite reached the end of my rope, but I was close enough.

BY THE END of my first year at the Department of Education, I took a dim view of the prospects for blacks in America. I no longer thought that the Reagan administration could do anything that would be of any help to them. Most blacks, egged on by the news media, had long since written off Ronald Reagan as a racist—a falsehood that hurt him deeply—and any black misguided enough to accept a job in the Reagan administration was automatically branded an Uncle Tom. It irked me to no end that civil-rights leaders and black Democratic operatives could casually refer to President Reagan's racism as if it were a universally acknowledged fact, and it made me no less angry when the white media repeated these slanders against him (and, by extension, those of us who worked in his administration) without producing a single shred of evidence to back them up. It seemed to me that white reporters felt obliged to cast racial aspersions on the president in order to prove their own sensitivity and moral superiority.

At the same time, those of us who had chosen to work for President Reagan found it hard to shake the nagging feeling that his aides didn't trust us. Though the president himself was always perfectly congenial and gave no sign of ever being anything other than wholly at ease with people of all colors, blacks never quite seemed to fit into the Reagan administration, except for the ones who worked for Vice President Bush. Too many political appointees appeared to me to be too preoccupied with celebrating their own ideological credentials to pay attention to the needs of blacks. We hadn't voted for him, so why should they bother with us? Some white conservatives appeared to take it for granted that no black could be truly conservative, which was at least as ugly as the liberal belief that no sane black with a clear conscience had any business working for Ronald Reagan. This left us in a no-man's-land where neither side thought we belonged, albeit for different reasons. It also led to our being excluded from the administration's policy-making process, an omission that caused the administration to commit a series of blunders which ensured that most blacks would never take its goodwill seriously.

The worst of these blunders was the Bob Jones University case. Bob Jones was a Christian college and seminary in South Carolina that maintained a number of racially discriminatory policies, including a ban on interracial dating among its students. (The ban remained in force until 2000.) The Internal Revenue Service revoked the university's tax-exempt status because of these policies. The school's appeal of the IRS decision was pending in the Supreme Court when, in 1982, the Justice Department, which had originally asked the Court to hear the case, abruptly decided not to pursue it any further. The Supreme Court promptly appointed William Coleman, one of the finest lawyers in the country, to argue the government's position in an amicus curiae brief, and subsequently ruled against the university. Like most blacks who worked in the Reagan administration, I supported the original IRS decision and was shocked when the Justice Department backed down and let the university off the legal hook. This made us feel like nonentities within the administration, and exposed us to scorn and ridicule from without. Yet my white colleagues found it impossible to understand why we were so upset. I explained to one of them that the case had destroyed the administration's credibility on race relations, to which he coolly replied that the story would be forgotten after a few days. The breakup of AT&T, he pointed out, was announced simultaneously, and that was what people would remember. I told him he had it exactly wrong.

I came close to resigning from the Department of Education over the Bob Jones case. The only reason I stayed on was because I still believed in the Reagan administration's commitment to limiting the role of the federal government in the lives of blacks (and everyone else). Aside from the oft-demonstrated incapacity of big government to solve our problems, I feared that the unintended effects of social-engineering policies like urban renewal would be at least as bad as the problems themselves. I no longer hoped to have any positive impact on race relations or on black attitudes toward the administration, but at least there was a chance that I could help solve a few specific problems. Above all I wanted to do what I could to keep historically black colleges from being thoughtlessly swept away in the rush toward integration. One day I mentioned to a senior career staffer that I didn't understand why there was so little focus on the educational role of the black colleges. He snapped back that they had no right to exist. His statement floored me, though in the next instant I realized that it was implicit in the position taken by the NAACP Legal Defense and Education Fund and other groups, which were obsessed with the racial composition of white colleges to the virtual exclusion of all other considerations. At no time in our discussions with these groups did the education of black students take center stage. All they seemed to care about were the numbers.

It distressed me that so few black college presidents were willing to speak out against the Legal Defense Fund's approach, or defend our efforts to help them survive. They, too, were unwilling to break ranks, even though their very existence was at stake. Two notable exceptions to this rule of silence were James Cheek of Howard University and Ernest Holloway of Langston University. President Cheek openly embraced the Reagan administration and its black members, and worked with us on a variety of issues to improve the university. Dr. Holloway was even more forthright; there was a real possibility at that time that Langston would be closed, and he was prepared to do anything in order to save it from extinction. I grew fond of him, and admired his willingness to speak frankly about the unintended effects of racial integration at a time when few other black leaders would discuss such sensitive matters, frankly or otherwise.

IN FEBRUARY 1982, Pendleton James, President Reagan's director of presidential personnel, asked me to come see him. The administration was looking for a new chairman to head the Equal Employment Opportunity Commission. The president's first nominee, William Bell, a black lawyer from Detroit, had been voted down by the Senate, largely because of strong opposition from civil-rights groups. I said I felt sorry for Bell, but that under the circumstances, I couldn't think of anyone who would want the job, nor would I advise any of my friends to take it. I said that I was having a hard enough time at Education because of the public's perception of the Reagan administration's racial attitudes, and couldn't see any reason to jump from the frying pan into the fire. "Are there any circumstances under which you'd agree to run EEOC?" he asked. I replied that anyone who took the job would have to be given total independence: no pressure to hire unqualified political appointees, no pressure to pursue an ideological agenda, and no attempts to cut the agency budget indiscriminately. "But all this is purely hypothetical," I added. "I don't want the job. In fact, I'm thinking of leaving the administration." Pen ended our conversation by asking what my answer would be if the president himself asked me to take the job. I hesitated, then reluctantly admitted that I'd have to say yes.

In fact I knew a little something about the Equal Employment Opportunity Commission, and I didn't like what I knew. Shortly after President Reagan was elected in 1980, Jay Parker had been put in charge of the EEOC transition team. He'd asked me to help out, and I was glad to oblige; I'd never been able to understand how an agency that had a mandate to assure equal employment opportunity for all citizens had gotten sidetracked into pushing race-conscious employment policies. So at Jay's request I paid a visit to the agency's Foggy Bottom headquarters, where I learned that it was a horrible place to work, a converted hotel that was filthy and poorly maintained. I pitied the people who had to labor in such unpleasant conditions. Why on earth would I now want to join them?

Pen James called me again on February 12, Lincoln's birthday. This time he skipped the usual pleasantries. "I just got back from the Oval Office," he said brusquely. "The president wants you to go over to EEOC as chairman." I said nothing. "Will you still do it?" he asked. I took a very deep breath, then said yes. "The President wants EEOC off the front pages of the newspapers," he added. That was the only order I ever received from President Reagan, then or later. Pen told me that he'd issue a press release that afternoon announcing my appointment, thanked me for helping the administration, and hung up. I sat in silence for a moment, wondering what I'd gotten myself into. Given my heterodox views and the Reagan administration's poor reputation on civil rights, I had a pretty good idea of what to expect from civil rights leaders and the media: first skepticism, then open hostility. For a fleeting moment I couldn't help but feel the urge to run, to go back home to Georgia and the uncomplicated life I had left behind so long ago. Resigning myself to my fate, I called Diane into the office and told her what I'd done. Then I summoned the rest of my personal staff and announced that I was going to EEOC. Anita Hill immediately said that she wanted to go with me. I said I'd think about it, reminding her that her position at Education was safe, thanks to a new collective-bargaining agreement that had given career attorneys the same job protections as career civil servants. "You're a rising star," she replied. "I want to go with you." I brushed off her description and said once again that I'd think about it.

The dread didn't start to set in until after I got home that night. I was still mired in debt, and had never stopped brooding about the damage I had done to my family. Not only had I hurt Kathy and Jamal, but my decision to leave them had strained my already shaky relationship with Daddy. I'd been careful to warn Aunt Tina that I couldn't bear to hear him criticize me about so painful a subject. He never did, always skirting the subject gingerly—but I knew how he felt. He'd always liked Kathy, and it was all too easy for me to imagine what he must have thought of my decision to walk out on her, leaving his beloved great-grandson behind. "You'll probably end up like your no-good daddy or those other no-good Pinpoint Negroes," he had told me on the morning he threw me out of his house, and his terrible words still burned in my memory a decade and a half later. Had Daddy been right after all? I poured myself a large glass of Scotch and Drambuie over ice and downed it greedily, alone with my thoughts and afraid of what lay ahead.

7

"SON, STAND UP"

The sight of my name on the front page of the *Washington Post* the next morning sent me into something close to panic, but the article itself, announcing my nomination to lead EEOC, was fairly ordinary. Only one detail stood out: a Hispanic critic claimed that I'd been "insensitive" to Hispanics during my time at the Department of Education. Over time this charge would blossom into the allegation that I was prejudiced against Hispanics as a result of an incident that supposedly occurred in law school. It was, of course, a fabrication, but there wasn't anything I could do to disprove it, since the original source of the slander had been some anonymous person who didn't have the courage to put a name to his claim. As a result it became part of the public record, available forever after to be regurgitated by anyone who disapproved of me. Cynics had long told me that perception was more important than reality in Washington, and stories like this made me fear that they were right.

Secretary Bell asked me to suggest a replacement for myself. I knew he'd need somebody who was bright and independent, so I recommended Harry Singleton, a Yale classmate (and my son's godfather) who was then a deputy assistant secretary at the Department of Commerce. Harry agreed to take the job, and he also agreed to keep all of my political appointees—as well as Anita Hill. But she told him the same thing she'd said to me: I was a rising star, and she wanted to follow me to EEOC. I wasn't pleased that she'd turned Harry down so tactlessly and started to have second thoughts about considering her for a position, but Gil Hardy intervened yet again. I said that I needed someone with experience in the field of employment discrimination, but Gil insisted that I should "give a sister a chance," attributing her behavior to naivete and the fact that she didn't know Harry.

For the moment I put aside the problem of what to do with Anita and concentrated on the immediate matter of getting myself confirmed by the U.S. Senate. It proved easier than I'd expected. The Senate confirmed my nomination by a unanimous voice vote on May 6, and a week and a half later I was at work. Even before Orrin Hatch, the chairman of the Senate committee that oversaw EEOC, warned me that the agency was in disarray, I'd taken the precautionary step of asking an Education Department career manager named John Seal to go over to EEOC and assess the situation there while I was waiting to be confirmed. John's report was full of bad news. EEOC, he said, hadn't reconciled its books in ten years and lacked an accurate accounting system, meaning that no one knew how much money the agency had on hand. Bills were paid so late that vendors would only do business with EEOC on a cash basis.

As John ticked off one piece of trouble after another, I slumped in my chair. It was my first day on the job, and I was already feeling like the fall guy. Eleanor Holmes Norton, my predecessor, was a civil-rights icon who'd been praised for her management of the agency. I was a nobody—too young at thirty-three, too green, too conservative, and a Reagan appointee. Who would believe me if I said that EEOC was on the verge of collapse? Yet on the very day I arrived, the General Accounting Office released a report condemning the agency's record keeping, and it was widely known inside the federal government that EEOC had great difficulty in enforcing the equal–employment opportunity laws.

Nor, I quickly learned, was there any chance of fixing things fast. During my first full fiscal year at EEOC, the agency closed out more than 74,000 cases of alleged employment discrimination, a new record. Some thought that was good news, but I knew it really meant that too many cases were being settled without adequate investigation. Instead of continuing to generate numbers that would look good in the press, we had to start concentrating on serious, effective enforcement—and the two goals were incompatible, at least at first. We'd have to build our wagon while we rode in it, and while we were doing so, our numbers would go down. Therein lay my dilemma: how could I fix an agency that its critics wouldn't admit was broken, especially when the only possible solutions were slow, painful, and certain to provoke outraged protests from the same people who'd caused them in the first place?

The deeper I dug, the more disturbed I became. One of the first things I did after reporting for work was pay courtesy visits to the other commissioners. EEOC was and is bipartisan: it consists of two Republican commissioners and two Democrats, plus a fifth commissioner, designated chairman by the president, who is usually a member of the President's political party. Two of my fellow commissioners had wanted my job, and one obviously knew little about the agency and the equal–employment opportunity laws we enforced. I knew I'd find no allies there, but at least I could do something about my executive director, the top manager in the agency. Early in the confirmation process, I'd met with all of EEOC's top political appointees and told them not to make any important decisions prior to my arrival. The executive director responded by coming to my office at the Department of Education at the end of a long day and asking me to accompany him to his office at EEOC, where he delivered a prepared presentation on his plans to slash the agency's budget. I declined to go along with his scheme, and he immediately complained to the White House. That told me everything I needed to know about his loyalty—and his judgment. I insisted on his immediate resignation as soon as I got to EEOC. I knew I had to seize the reins immediately, so I assumed the position of executive director myself, figuring that if I was going to take the heat for EEOC, I might as well run the whole show.

When the *Washington Post* found out what I'd done, it wrote up my decision with an unexpected twinkle: "In an extension of the premise that if you want a job done right, you should do it yourself, the chairman of the Equal Employment Opportunity Commission has also been acting as the agency's executive director." That gave me a laugh—one of the few I had in 1982. It was also one of the very few accurate pieces about me to appear in the press during my eight-year tenure at EEOC. I didn't blame Al Sweeney, who ran our office of public affairs. He was making the best of a bad situation; I liked and trusted him. But he couldn't write the stories himself, and those who did often seemed to be pursuing other agendas. On his advice, I gave my first interview to Ernie Holzendorf, a young black reporter for the *New York Times.* I told Holzendorf that I'd do all I could to open the doors of employment opportunities for all citizens, especially minorities, but that EEOC had no power to prepare anyone to take advantage of those opportunities. To my surprise, what appeared in the *Times* was an article consisting mainly of quotes from people who disapproved of my views, none of whom knew me. Sweeney explained that Holzendorf's editors had told Holzendorf that I was "controversial" and made him rewrite his piece to reflect this supposed fact. "How can I be controversial?" I asked him. "I've only been here for a month!"

It also surprised me that the *Times* made no mention of the hard work we were doing to address the agency's nearly overwhelming problems, but I soon got used to that kind of omission. It's easier for reporters to pontificate than to try to fathom the complex inner workings of a government agency, as I learned to my detriment when Carl Rowan, a black syndicated columnist, wrote a piece called "Wrong Man at EEOC" based solely on the Holzendorf profile. Rowan hadn't bothered to talk to me, so Al Sweeney suggested that I have lunch with him. Rowan promised at the lunch not to write about me again without contacting me first—a promise he never bothered to keep.

Not only was EEOC a manager's nightmare, but the building in which we were housed was little short of squalid. The carpets were flea ridden and the windows were covered with mold. Shortly after my arrival, I noticed that someone had spilled coffee in one of the hallways. The puddle was still there several days later, somewhat reduced by evaporation but otherwise undisturbed. On another occasion I ran into a staffer in the hall who was carrying a package of copying paper in one hand and some work materials in another. I asked him what he was doing. He said he was looking for a copy machine that worked, adding that paper was so hard to find at EEOC that he kept his personal supply locked up in a desk drawer for safekeeping.

These were outward signs of the deeper disorder with which the agency was afflicted.* In the fiscal year prior to my arrival, EEOC had to return more than $10 million of its $140 million budget to the Treasury Department because of poor financial management. Because of this, it was forced to lay off employees and put off paying its vendors. I knew it would be impossible to do any real work until I got a handle on the agency's finances and learned how to use our resources efficiently, neither of which would be easy. Throughout

* Another, even more ludicrous one was the cash award given to a marginal employee for having put together an EEOC phone directory—a routine task in any properly run agency.

my first year, financial data still had to be recorded on old-fashioned paper punch cards before they could be entered into our ancient computers. It was an appallingly labor-intensive job, slow, tedious, and prone to error, in part because the computer room wasn't climate controlled. I still cringe when I remember the Friday we stored a batch of newly punched cards in the computer room. It rained that weekend, and when the staff returned on Monday and tried to feed the cards into the computer, it turned out that they'd become too damp to be read. But despite such demoralizing mishaps, we managed to close the books and put a reliable accounting system in place by the end of my first year. (We also installed a climate-control system.) In time we automated the entire financial system, increasing its accuracy to the point where EEOC regularly used over 99 percent of its appropriated funds and acquired a perfect record for paying its bills on time, an achievement that qualified us for substantial discounts on goods and services. All of this allowed us to make long-term improvements in our operations without receiving an extra dime from Congress. We had to. Congress never passed any appropriations to rebuild EEOC and routinely gave us less money than the President requested—which didn't stop its members from criticizing our deficiencies.

I had to enlist the support of EEOC's senior managers before I could do any of these things, so I spent several days meeting with them after coming to the agency. These meetings were inevitably awkward, both because of the public's negative perception of the Reagan administration's civil rights record and because I was much younger than most of the people working under me. I also had to do something about Anita Hill, who'd been pestering Anna Jenkins, my interim secretary, as had Gil. I reluctantly brought her aboard, and the first thing she did was claim the largest office in my suite. She had no experience with employment law, so I also asked two outstanding young career employees, Allyson Duncan and Bill Ng, to join my personal staff. Their political views didn't matter to me. They knew the law, and that was what I needed.

I found it easier to deal with EEOC's financial problems than to straighten out my personal budget. In May, just as I arrived at EEOC, I had to find a way to pay the first installment of Jamal's tuition, which was due the following month. I had no money and no credit, but I got a miraculous break at the last minute: I received a notice in the mail informing me that my car, a Fiat, had been recalled because of a rust problem. Apparently Ralph Nader, the longtime consumer activist and self-appointed left-wing gadfly, was responsible for the recall, an irony that wasn't lost on me. Fiat paid me more than $2,000, enough to settle the car loan and take care of Jamal's tuition bill. The catch was that I didn't have a car anymore, but that was fine with me, as I usually stuck close to my apartment, doing chores or clearing up EEOC paperwork in the evenings and exercising every morning. Since I lived in southwest Washington, I ran along the Mall before dawn, pausing at the Lincoln Memorial to stretch, then returning to my apartment to change clothes and meet Mr. Randle, the driver who took me to and from work in the agency's car.* President Lincoln, the Great Emancipator, had been a hero for southern blacks during my youth. Now he took on a more personal role in my own life as I gazed up at his handsome monument at the crack of dawn or, when I ran in the evening, bathed in light against the dark night sky.

I spent many hours talking to Phyllis Berry, who had come to the agency with me from the Department of Education, during my first

---

* James Randle—I always called him *Mister* Randle—was a World War II vet from Mississippi who had been driving for the federal government since the early fifties. He was a man of impeccable discretion who never repeated anything he overheard in the agency car. (He wouldn't even say where he'd taken any of my predecessors.) Mr. Randle became a wise, loyal confidant, and I could never have come to understand the agency as quickly as I did without his help. I came to think of him more as a senior adviser than as a chauffeur.

weeks at EEOC. Phyllis sternly warned me that some of my enemies would stop at nothing to destroy me, and suggested that I do myself a favor and stop drinking. "Loose lips sink ships," she said. Her words lingered in my mind, like a bad dream. Then I went out one Friday evening with Thelma Duggin, an old friend from the Hill who had since become a staunch ally at the White House. I drank too much and ate too little, and by the time I finally stumbled home to my sad little apartment, I was exhausted. The room was spinning and I felt queasy as I fell asleep on the old mattress I used as a bed. (Three years went by before I bought a bed of my own.) I had a splitting headache the next day. I opened the refrigerator to get something to eat. It was empty—except for two cans of Busch. I thought of how long it took me to clear my head the morning after I'd had even a few drinks, and vowed that those two cans of beer would be the last alcohol I ever drank. I drew myself a hot bath and downed them slowly as I sat in it. I haven't had a drink since.

WHEN I ARRIVED at EEOC, I inherited two major cases involving General Motors and Sears. Both were the result of broad class-action suits that had been filed on the basis of statistical analyses of hiring and promotion patterns. My predecessor had charged Sears with failing to hire or promote enough blacks and women to the company's more lucrative commission-sales jobs. In fact, though, there were no actual job applicants or employees alleging discrimination; the charges were based solely on the fact that these groups were numerically underrepresented relative to their presence in the population.

The Sears case was the most expensive item in EEOC's discretionary budget. It was also a no-win situation, for the costs of litigation were eating up the funds we needed in order to put the rest of the agency on the right track. Yet as long as it was in court, I felt obliged to provide the funds that our attorneys needed to litigate it properly, most of which went to outside experts who could provide the statistical analyses that underpinned the case. Some of my other managers, seeing their budgets squeezed to fund the case, complained bitterly that the only people who stood to benefit from it were the expert witnesses and Sears's lawyers. I explored the possibility of settling the case, but was told that Sears's chairman and CEO felt so strongly that the case was unjust that he intended not only to defend it to the bitter end but to push for his legal fees. He prevailed, but the company was awarded only a portion of the legal fees, which reportedly totaled more than $20 million. EEOC had also spent millions prosecuting the case. I was right: nobody won.

The General Motors case proceeded very differently, though the allegations were similar. GM adamantly denied all charges of discrimination, but it preferred not to waste millions of dollars defending itself in court, and within a month of my arrival at EEOC, Abe Venable, one of the company's executives, approached me to discuss the possibility of settling the case before it went to court. I told him that it was still being investigated, and that I'd need a full year to get our management crises under control before we could discuss a settlement. Abe got back in touch with me almost exactly a year later, in the spring of 1983. The last thing I wanted was to secure a financial settlement, then go looking for putative victims on whom to bestow it. It occurred to me that we might be able to use the money to endow scholarships for the children of GM workers or other young people to whom it would make a difference, so I asked Allyson Duncan to work with Jim Finney, one of my senior career attorneys, on the settlement, instructing them to think outside the box about how to craft an agreement that would help people instead of merely looking good.

When Jim, Allyson, and GM's representatives finally reached an agreement in the fall of 1983, the results far exceeded my expectations. The settlement included a payment of more than $40 million, of which more than $10 million was to be distributed to various schools, colleges, and universities, for permanent endowments to assist deserving students, preferably (but not limited to) the children of minority and female employees of GM. Many of the endowments were established at historically black colleges and universities such as Florida A&M University and Savannah State College, though others went to schools such as the University of Pittsburgh and Yale.* We also included grants to such organizations as the Society of Women Engineers, the National Urban League, Inroads, the National Hispanic Scholarship Fund, and the Minority Leaders Fellowship Program of the Washington Center. Many of the presidents of the historically black colleges who profited from the settlement were surprisingly reluctant to express their gratitude for EEOC's efforts to help their students at their institutions. Some pointedly referred to the scholarships as "GM scholarships," although they were in fact known as EEOC/GM Scholarships. In addition, many civil-rights groups were openly dismissive of the settlement, as were congressional Democrats who chimed in with an oversight hearing questioning the legitimacy of the agreement. One congressman, Bill Clay of Missouri, grilled me on the terms, apparently offended that a member of the Reagan administration had dared to try something new. (To his credit, he later dropped the matter.)

The near-total unwillingness of Democratic congressmen to acknowledge the significance of what we'd done may seem remarkable

---

* The only university official to express concerns about the arrangement, as it happened, was Bart Giamatti, the president of Yale, who called me to explain that he had just gone through the delicate process of eliminating scholarships that were designated for white or male students and didn't want to now establish a race-based scholarship there. I assured him that the scholarships did not exclude any race or sex, after which he took the money graciously.

in retrospect, but the longer I worked at EEOC, the more acutely aware I became of the difference between what happened inside the agency and what the rest of the world thought was happening. No matter what we did to improve our operations—and we did plenty—I continued to be portrayed as controversial in the press and treated with disdain by Democratic congressional staffers who neither understood nor cared how EEOC worked. When I reassigned staff to match workforce with workload, I was "gutting" the agency; when I declined to pursue fruitless lawsuits, I was "cutting back on enforcement." What's more, these staffers had the ear of members of Congress who were too busy to do anything but take their false claims at face value. This gave them the power to make my life miserable, and they used it.

Since Republicans had taken control of the Senate when President Reagan was elected, the Republican staffers on our Senate oversight committee were cooperative and helpful. Senator Hatch also understood what I was trying to do—he'd never had any illusions about my predecessor's managerial abilities—and he would always be there when I needed support. But my relationship with the Democratic House's oversight committee, Education and Labor, was sour from the word go. The employment subcommittee, chaired by Congressman Gus Hawkins, a Democrat from California, held hearings not long after I arrived at the agency, and Congressman Hawkins started off by accusing me of trying to destroy the agency. That was news to me. All I wanted to do was make it work, and I never understood how any sensible person could possibly think that a black man from the Jim Crow South would devote himself to destroying an agency that existed to enforce equal-employment laws. At first I thought that we merely disagreed about the best way to reach a common goal, but I came to realize that Congressman Hawkins and his staff assumed that I was a bad person simply by virtue of having been appointed by Ronald Reagan.

Perhaps I should have been more realistic about the ways of Washington. When I arrived in 1979, I'd naively supposed that I was joining a community of people who had chosen to work in politics to do some good in our society. No less naively, I'd expected to find a certain solidarity or kinship—a shared sense of purpose—among the blacks I met there. I accepted it as a given that politicians had to say and do certain things in public in order to keep their constituents happy, but somehow I had the idea that blacks in Washington would at least be able to talk honestly in private about our problems, then work together to resolve them. But I felt no common cause with the black Democratic congressional staffers, or with most of the reporters who wrote about EEOC during my years there. So far as I could tell, we didn't care about the same things, and sometimes I wondered whether the reporters cared about anything. One reporter told me that good news about civil rights simply wasn't "newsworthy" during the Reagan years. As far as I was concerned, that said it all.

Of course the Reagan administration was partly to blame for this wretched state of affairs. The aggressive, tone-deaf manner in which the administration made some of its early civil-rights decisions had helped earn it a reputation for racial animus that no subsequent achievements, however impressive, could alter. President Reagan sensed this, and was disturbed by it. One day I came back from lunch to be told by Diane that the president had called while I was out. I assumed it was a prank, but she called the White House and—to my horror—put the president on hold while I came running to the phone. I was stunned to hear his voice and amazed to hear what he had to say. He'd called to thank me for my work at EEOC, but he also told me he was hurt that he was being called a racist and was mystified by his inability to persuade blacks that he wasn't. He'd never believed in discrimination or segregation, the president assured me. The sincerity in his voice was unmistakable. I was embarrassed that he was taking time from more urgent matters to talk to me, and troubled that he'd been stung by the unfair criticism. I, too, was disappointed by our inability to persuade blacks of his good intentions, but I did my best to buck him up.

Not everyone wrote unsympathetically about EEOC during my time there. Nancy Montweiler of the *Daily Labor Report* was the most honest and conscientious of the reporters who covered us. Instead of viewing the agency's activities in a purely political light, she worked hard to understand what we were doing, and while she was as aggressive as any reporter in town, she didn't write in the nasty "gotcha" style of so many of her colleagues. As for the columnists who wrote about EEOC, the most comprehending and scrupulous was Warren Brookes of the *Detroit News*. He always called me in search of relevant data whenever he was writing a piece, though he never took what I told him on blind faith. On more than a few occasions, he pointed out discrepancies in the information I gave him and asked for explanations.

My closest relationship with a journalist, though, was with Juan Williams of the *Washington Post*. I didn't blame him for quoting me about my sister in that first story he'd written about me in 1980. I'd come to believe that he was a decent man looking for the truth, and I appreciated the courage that led him to avoid the reflexively accusatory approach of so many of his colleagues. Juan made some mistakes in his coverage of EEOC, but he tried to understand the agency and its activities, and I respected him for that. We would speak frequently and frankly throughout the next few years, not merely for professional reasons but also as black men and fathers.

I was honest with Juan about my view of the race debate, which I found increasingly exasperating. I didn't care which schools blacks attended, so long as they received a good education. As I told another reporter, "I think segregation is bad, I think it's wrong, it's immoral, I'd fight against it with every breath in my body—but you don't need to sit next to a white person to learn how to read and write." Nor did it matter to me if certain neighborhoods were predominantly white or black, so long as they were safe and blacks could freely choose to live in either. I was sick and tired of the theories and statistics that had come to dominate the discourse on both sides of the political fence. What mattered to me were individuals and their problems, but most of the people I met in Washington, Republicans and Democrats alike, seemed hell-bent on winning arguments instead of solving those problems. Juan understood my pent-up frustration with this self-serving, irrelevant debate, a constant struggle that never seemed to change anything for the people who needed our help.

I also got along well with Elaine Jones of the NAACP Legal Defense and Education Fund, to whom Lani Guinier had introduced me while I was at Yale. Elaine and I often disagreed, but there was never anything personal about it, and she treated me with the utmost professionalism, unlike some of the other representatives from the major civil rights organizations whom I met after going to EEOC. I lunched with a group of them early in my tenure, and they told me that they wanted to maintain a "working relationship" with the agency. That was fine with me, I said, encouraging them to participate in the public comment period of our rule-making process. I was brusquely informed that they wanted to be "at the table" right from the start, the same way they'd been during the Carter administration, and that they'd make my life difficult if I failed to cooperate. Once again I was naive: it astonished me that such advocacy groups took for granted their right to be granted access to the government's policy-making process in advance of the rest of the public. What would they have said if I'd extended the same access to business-advocacy groups? I grudgingly admired their gall, but I didn't care for the arrogance with which they made their demands, and I declined to cooperate, little knowing that it wouldn't matter in the least what I said or did. Democratic congressional staffers and a few disgruntled EEOC staffers invariably told them whatever they wanted to know whenever they wanted to know it.

WOMEN EMPLOYED WAS one of the groups that pressed for access to our internal documents. Nancy Kreiter, the shrewd, persistent research director, seemed to think that we weren't pursuing a sufficient number of "equal pay" cases, so she routinely asked for internal agency documents in the hope of bolstering her claim that we were weak on enforcement. I refused her requests, but she took her case to Congress and the press, which worked in tandem to help her get her way; by the end of my tenure, EEOC was required to provide her with quarterly reports. As a gesture of goodwill during my first year on the job, I flew to Chicago on March 30, 1983, to appear at one of the group's meetings. About a hundred mostly white women showed up. They gave every impression of being successful, and judging by the questions they asked me, they were smart and sophisticated as well. Yet I couldn't understand how angry they seemed to be about their lot in life. How could these well-off white women be more bitter than the poor blacks and Hispanics with whom I met regularly at EEOC?

By then Myers was working in Chicago for Sheraton Hotels, and I always visited him and his family whenever my work took me there. He came to pick me up after the meeting, accompanied by a coworker. My battles at EEOC and my anxieties about my personal life had drained me of what little energy I had, and I'd been looking forward to spending a quiet evening at home with Myers, Dora, and my little niece, Kimberly. But as soon as he dropped his friend off, he said abruptly, "Your daddy died."

"You mean C?" I asked.

"No, Myers Anderson—the only daddy you ever had." His words didn't sink in at first. "He had a stroke coming out of the field today.

Cousin Peasy took him to the hospital, but he died on the way. Leola tried to call you before you left Washington, but you'd already gone."

At last I understood, and it was as if he'd slapped me across the face. Daddy was lying dead in Savannah while I'd been talking about politics and policy to a roomful of angry white women in Chicago.

In an instant my mind snapped back to our last meeting. I'd flown down to Savannah the month before to attend my mother's wedding. Daddy had stubbornly refused to go, partly because he didn't think Leola should get married again and partly because Aunt Tina was in the hospital, being treated for heart problems. We'd spoken in the waiting area after I visited with her; it was a wonderful talk, the best we'd ever had. We agreed that the reason the two of us always had such a hard time getting along was because we were so much alike. Afterward we embraced, the first and only time in our lives that we did so. Might it be possible that he and I were finally growing closer? The embrace, I felt, had brought our battles to an end at long last, and it had also whetted my appetite for the truly intimate relationship that we'd never had. I'd returned to Washington full of hope—but now it was too late.

Myers took me home for a quick visit with Dora, then drove me to an airport hotel so that I could catch the first flight back to Washington the next morning, then go from there to Savannah. I couldn't sleep. All I could do was think of the things I wished I'd done differently. Why had I spent so much time arguing with Daddy instead of trying to understand him better? Why hadn't I moved back to Savannah to be near him instead of spending the last years of his life on thankless jobs in Washington? What a waste! At least we'd made a start on the process of reconciliation, but I hated myself for having succumbed in college to radical ideologies that had pulled us so far apart that we only came together at the last possible moment, bridging the gulf of mistrust with a single, tentative embrace. I would never be able to tell him how right he'd been, or how much I admired and loved him, or that it had been immaturity and false pride that kept me from forgiving him for the hardness with which he had treated Myers and me.

I flew to Washington the next day to make the arrangements necessary so that EEOC could carry on in my absence. The world around me seemed unreal and irrelevant. I half expected to get a call from Myers telling me that he'd only been playing a cruel joke on me, that Daddy was still alive. But I knew better. I'd talked to Leola on the phone, and she'd told me the details of his death. Daddy, she said, had been fighting off a bad cold. Instead of going to bed, he'd undoubtedly dosed himself with 666 Cold Preparation, an old-fashioned patent medicine popular in the Deep South, then spent the morning working in the field. I remembered how he'd always doubled the dosage whenever he was sick, thinking that it would help him get well more quickly. He'd also been taking a large number of prescription medications, and Leola suspected that the combination of drugs had sent his blood pressure skyrocketing, triggering a massive stroke. Eugene Osborne—our Cousin Peasy—had driven him to the hospital as quickly as possible. He complained on the way about being unable to see, then asked Cousin Peasy to take care of Aunt Tina. Those were his last words. He was pronounced dead on arrival at St. Joseph's Hospital in Savannah, two days before his seventy-sixth birthday and just shy of his fiftieth wedding anniversary.

After I got to Savannah, my mother and I went to the funeral home to view his body. It resembled him but was not like him: death had robbed him of the soul and spirit that had filled our lives. I touched his cold hand, still unable to fully grasp what had happened. He'd been larger than life, and in my heart I'd always supposed that he would prove to be larger than death as well. I thought my spirits couldn't sink any lower. Then I drove out to the farm to see Aunt Tina. By then she was only a shadow of the beautiful woman who had raised me. She was visibly weakened and unsteady on her feet, and her face was drawn and gaunt. Once her full cheeks had smoothed the lines of her face and made it soft and pretty. Now her high cheekbones stood out from the rest of her grief-stricken features. Only her still-perfect teeth remained to remind me of how she'd looked when I was a boy.

Kathy and Jamal, who had just turned ten, flew down for the funeral. Even after I left her, Kathy never lost her affection for Daddy, and her dignity and sincerity touched me deeply. No sooner did the service begin than I started to weep shamelessly and uncontrollably, something I'd never done in public. I wept for the man I revered, for things done and undone, for the unrequited love we shared for each other; I wept at the thought of his hard life as an uneducated black man in the South, a life that had ended too soon. Myers tried to console me, and Jamal patted me on the back, but the only thing that could have eased my pain was the touch of the man we were laying to rest. I wept beyond tears, slipping into the barren, rhythmic heaves of a body seeking something more.

Daddy had bought a pair of burial plots at Palmyra Baptist Church in the Sunbury area of Liberty County, one of two black Baptist churches close to the farm. (It bore the name of the nearby plantation from which my family had come.) The funeral procession traveled along Highway 17, the road we'd taken during our many trips to and from the farm. Glimpses of our past life flashed before me as I stared numbly at the familiar scenery. I remembered Daddy driving the Ford pickup, with Myers and me tucked in between him and Aunt Tina. Then I looked out the window and saw Mr. Sam Williams, his lifelong friend and business partner, standing beside the road to wave good-bye. The burial service itself was brief and final. Neither Aunt Tina nor Myers cried, and even my mother, who had never been emotionally strong, somehow managed to keep hold of herself. It was I who had no control over my grief.

When it was all over, I went back to work in Washington, where the controversies and problems seemed to mount faster than I could address them. I felt empty and lonely—and uncharacteristically vulnerable, though I tried to deny it. No matter how bad things got, and even when we were at our most distant, I'd always been able to look to Daddy as a living example of strength and fortitude. At one of our last meetings, I'd complained to him about how badly I was being treated because of my views. "Son, you have to stand up for what you believe in," he said. It was just that simple, for it was just what Daddy had done his whole life. Those were the words I needed to hear, and he was the man from whom I needed to hear them. Now he would never say them to me again.

I went back to Savannah a few weeks after Daddy's funeral to check on Aunt Tina. She had taken to her bed, but when she saw me, her eyes widened with love. I squeezed her hand and kissed her cheek, shocked by how frail she looked. Her tiny frame was barely visible under the bedspread. Yet she never cried and never complained. Later that evening I was talking with relatives in the living room when one of my cousins came running in, saying that Aunt Tina was unable to move. I rushed into the bedroom and found her limp and still. One of her eyelids was drooping. I picked her up and carried her to my rental car, put her in the backseat with Leola and one of my other cousins, and drove to St. Joseph's Hospital more slowly than I wanted but as fast as I dared. As soon as we got there, the emergency-room attendants confirmed what I already knew: Aunt Tina had suffered a stroke. They called her doctor, who said to our chagrin that there was no reason for him to come in or for us to stay the night. She was resting comfortably, he said, and there was nothing more to be done at the moment. He examined her the next morning and told us that she was unlikely to recover. Perhaps, I thought, she'd lost the will to live.

I called Myers in Chicago and told him to come home at once. He arrived on Sunday and went straight to the hospital. Aunt Tina was unable to speak, but he could tell that she recognized him. He sat with her for a little while, then went to visit our sister. Aunt Tina died as soon as he left, having clung to life just long enough to see her other son. I'd had to go back to Washington, but now I returned to Savannah to repeat the bitter ritual of grief and self-reproach. I hadn't thanked Aunt Tina enough for her sacrifices; I hadn't gone home often enough; I hadn't hugged her often enough, or told her how much I loved her. No sooner did Myers and I go home to the farm after the funeral than some of our relatives started fighting over the contents of the house, declaring that Aunt Tina would have wanted them to have this item or that. Part of me was disgusted by their greed, but I couldn't bring myself to care. Death had already stolen the only things in the house that mattered to me.

Losing Aunt Tina a month after Daddy was more painful than I could ever have imagined. How could I have let myself grow away from her, or from the man who, as Myers had truly said, was the only real father I'd ever had? The guilt with which I had been wracked ever since I left Kathy and Jamal was doubled and redoubled. As I suffered its torment, I came to understand that there was nothing I could do to assuage it but to embrace with humility the spiritual legacy of my grandparents. Long ago Daddy had said that their way of life and his unshakable belief in the redeeming power of work would be my inheritance. Now all I could do was unconditionally accept what he and Aunt Tina had left me. I would go back home—not to Savannah, but to them.

IT WAS EXCRUCIATINGLY difficult for me to concentrate on the day-to-day problems at EEOC that had come to seem so unimportant, especially since I had to fly back to Savannah several times that spring to settle my grandparents' affairs. But work waits for no man, and I went so far as to cut one trip short in order to put the finishing touches on our settlement agreement with General Motors, at the same time clearing up a minor task that I hoped would rid me of a growing nuisance. In the midst of my grief, Anita Hill had been nagging me to write her a letter of recommendation, and the sooner I did it, the sooner she'd be out of my hair.

At some point near the end of 1982, Chris Roggerson, my chief of staff, had filled me in on his interim performance evaluation of my personal staff. While most of the staffers had done well, Chris told me that Anita wasn't performing up to expectations and failed to finish her assignments on time. I hadn't realized that her work was so deficient, but I'd already noticed that she'd stopped coming to our morning meetings, evidently as the result of a quarrel with another staff member. Such quarrels were not uncommon—I'd also noticed Anita's rude attitude toward the other members of my staff—and it had been bothering me as well that she seemed far too interested in my social calendar. She regularly inquired about my after-hours activities and on more than one occasion had asked if she could accompany me to professional functions. Chris's unfavorable evaluation now caused her to become even more sullen and withdrawn.

Early in 1983 Chris asked to be moved to San Francisco to replace a district director who had resigned rather than accept a transfer. I hated to lose him, but he'd earned the right to ask for the job, and the least I could do was honor his request. I knew I needed to replace him with someone who had a strong background in equal-employment opportunity policy, and I thought at once of Allyson Duncan and Bill Ng. Neither one had asked to be promoted, though it was obvious that they were the most qualified candidates on my personal staff. Instead it was Anita who approached me about the job, telling me that she deserved it because she'd gone to Yale Law School. (Allyson had gone to Duke University, Bill to Boston College.) It would have been hard for her to come up with an argument less likely to sway me, and it confirmed my feeling that she wasn't cut out to be a supervisor.

I picked Allyson, a consummate professional whose work had been consistently outstanding. As soon as I announced her promotion, Anita stormed into my office and accused me of favoring Allyson because I liked light-skinned women. (Not only was Allyson a light-skinned black, but so was the woman I'd been dating for some time.) She insisted again that she was more qualified than Allyson, having gone to a more prestigious law school. I found her accusation, her attitude, and her reasoning equally irritating, and told her so. She replied that she'd start looking for another job, then stormed out of my office as abruptly as she had barged in. I called Gil Hardy to tell him what had happened, but once again he came to her defense and asked me to be patient with her.

Not long afterward, Elaine Jenkins, a well-known Republican whose husband Howard was the only black member of the National Labor Relations Board, asked me to be the luncheon speaker at an EEO seminar at Oral Roberts Law School in Tulsa. The timing was bad—Daddy had just died and I was still in shock—but Elaine insisted that I accept the invitation, and I reluctantly agreed. It occurred to me that this would be a good opportunity for Anita, who came from Oklahoma, to spend some time with her family and think through her options, so I suggested that she also participate in the seminar. This seemed to mollify her, and she flew out a few days before the seminar. I didn't want to spend the night in Tulsa, so I took an early-morning flight on the day of the luncheon and came back that same evening. As soon as I arrived, I was introduced to Charles Kothe, the dean of the law school. Dean Kothe told me that Anita had performed well at the seminar and asked my permission to approach her about joining his faculty. Seeing a chance to solve a problem of my own, I told him that she was looking for a job and might well appreciate the opportunity to go back home to Oklahoma. I added that she hadn't performed as well at EEOC as I'd hoped, but that this might be nothing more than a matter of immaturity.

Much to my relief, Anita accepted Dean Kothe's offer. All she needed, she said, was a formal letter of recommendation to complete the hiring process. I would have been glad to supply it, but the death of my grandparents had made it hard for me to cope with even the most important of my duties at EEOC, much less write letters of recommendation. I finally got the letter written and sent off, though my schedule prevented me from accepting an invitation from Sonya Jarvis, Anita's roommate, to attend her going-away party. Revealingly, EEOC staff had refused to throw a party for her. Anita asked me to speak at an EEO function in Tulsa three years later. I agreed, and shared one of the head tables with Anita, Dean Kothe (who by that time was working for me at EEOC), and his wife. Dean Kothe had planned to drive me to the airport the next morning, but Anita, who had come over for breakfast, insisted on taking me in her new car. She was excited about the car and seemed happy. I could tell that Dean Kothe liked her, and I was glad—and relieved—that everything seemed to be going well for her. She called me from time to time after that, but so far as I can remember, I never saw her again.

I sometimes wonder how I got through the summer of 1983 without falling apart. As we say in Georgia, I was lower than a snake's belly. I told Diane Holt that if I so much as tripped and fell, I didn't think I had it in me to stand up again. I meant it: I'd actually reached the point where I wondered whether there was any reason for me to go on. The mad thought of taking my own life fleetingly crossed my mind. Of course I didn't consider it seriously, if only because I knew I couldn't abandon Jamal as I had been abandoned by C. But I did ask myself whether I might do better to back away from my political beliefs. Life, I knew, would be so much easier if I went along with whatever was popular. What were my principles really worth to me? As I gazed out my office window at the Potomac River, the answer came instinctively: *They're worth my life.* I spoke the words out loud, knowing at once that they were true. My beliefs had come from the two people I had just buried. Though Daddy and Aunt Tina hadn't been my parents, I was their son. I vowed that day to live the rest of my life as a memorial to theirs. I had a reason to live—and to keep on standing up for what I thought was right.

Unfortunately, I couldn't use my principles to pay my bills, which continued to pile up. When I got my American Express bill in June, I saw that the trips I'd made to Savannah had cost more than $2,000. I didn't have it, so I couldn't pay it. Before long I started getting increasingly urgent phone calls at the office demanding that I pay my bill and warning me that my account would be canceled if I failed to comply. That was a serious threat: it was the only major credit card I had, and I used it mostly to settle my hotel bills when traveling on government business. I had no choice but to take out a high-interest consumer loan from Household Finance. But even after I paid the bill, American Express cut me off. From then on Diane had to book me into hotels that would accept cash. On one of my trips to Massachusetts to attend a meeting of the Holy Cross board of trustees, I tried to rent a car at the Boston airport with an old Sears credit card. The clerk at the Budget rental desk called the company, then told me that Sears had ordered him to destroy the card. He cut it up on the spot as I looked on in horror. I had to beg him to let me rent a car so that I could get to my meeting.

Things kept on going from bad to worse. Running EEOC was a Sisyphean struggle: every time we put out one fire, another one started. My bills piled up, often unopened. I was nearly evicted from my apartment more than once because I'd been late with the rent. Daddy and Aunt Tina were rarely far from my mind, and I cried uncontrollably each time I thought about them. I listened to a lot of blues and country songs that summer, often thinking that my life had become just like their mournful lyrics. I did all I could to put on a good face at the office, not always successfully. Daddy had told me that while life was hard, enough elbow grease would make it easier. Instead it kept on getting harder and harder. But he'd said something else as well: "Son, give out but don't give up. Get up every day and put one foot in front of the other." That was what he had always done, and so had Aunt Tina. If I wanted to live for them, I would have to do the same thing.

By then I had an even more important reason to keep on getting up every day. My divorce from Kathy would soon become final, and we'd agreed that Jamal would come to live with me in September. This meant that I had to find a larger apartment—and buy some furniture, too, something I'd never bothered to do. I found an inexpensive three-bedroom apartment in Hyattsville, Maryland, a few blocks outside the District of Columbia. The building was full of cockroaches and the walls were paper thin, but I couldn't afford anything better, especially after I went further into debt to buy beds, bureaus, desks, and a couple of chairs. Fortunately, Jamal proved to be both independent and dependable, and I soon found that I could trust him to take care of himself. At first he spent weekends with Kathy, but the two of us lived together full time after she moved back to New England in 1985. In our spare time we ran errands, took trips, played video games, listened to music, watched TV, and ate too much. Taking care of my son while holding down a demanding job was difficult, but it helped to ease my guilt and gave me another reason to go on living.

As we walked home from the grocery store one evening, I told Jamal that I hoped he'd someday be able to understand why I'd left his mother. I said that my decision to leave had had nothing to do with him; I solemnly promised that I'd never say anything negative about Kathy and that I'd do all I could to make his life less painful. He looked at me with quiet, seeking eyes. I could tell that my words had made an impression on him, but I still felt like crawling into a hole and crying. I knew that I'd hurt my own child, and that I could never repair the damage I'd done. We would simply have to endure. I could only hope that Jamal might learn to trust me again someday—but I also knew that I'd have to earn his trust. I didn't deserve it yet.

NOT ALL OF my problems at EEOC came from Democrats. The Justice Department's Office of Civil Rights had the broadest civil-rights enforcement authority in the federal government, as well as far more prestige than EEOC. Our authority to bring lawsuits was limited to employment discrimination in private companies that had fifteen or more employees, but the Justice Department's broader jurisdiction covered all aspects of racial discrimination, including the enforcement of anti-discrimination laws against state and local governments. Bradford Reynolds, the assistant attorney general for civil rights, didn't think consent decrees that included race-based numerical goals and timetables were consistent with the civil-rights laws, so it was his practice to challenge existing decrees that included them. The Justice Department had acted alone in these matters prior to my arrival at EEOC, but after the Bob Jones fiasco, Brad started meeting with me from time to time to coordinate our efforts, and we agreed to keep each other informed before taking any major positions in civil-rights cases.

Our agreement went out the window in January 1983 when I opened the *Post* and read that the Justice Department had decided to challenge the use of goals and timetables for hiring in the New Orleans police department. I was upset with Brad for blindsiding me, while my fellow EEOC commissioners were downright furious.

One of them actually threatened to go over to the Justice Department and punch him. I managed to calm everyone down by promising that we'd look into the possibility of preparing a brief that would support the use of goals and timetables. That wasn't my position, of course—I didn't like that kind of quota-based thinking any more than Brad did—but a majority of the commission favored it. Then, as always, I felt morally obligated to advocate our official position, even when it conflicted with my personal views.

I told Brad I was unhappy that he had acted without informing me and warned him that we would be filing a brief opposing the Justice Department's decision. Before it was completed, I was asked to meet with Attorney General William French Smith to tell him what gave EEOC the authority to take such an action. I argued that we were an independent agency and could do as we pleased, but the attorney general responded that our litigation authority was limited to the private sector and that Eleanor Holmes Norton, my predecessor in the Carter administration, had conceded as part of the government reorganization of 1978 that EEOC was an executive-branch agency.* While I hadn't been aware of that precedent, I still thought that we had a legal right to file our brief and I refused to back down. Another meeting was called, this time in Ed Meese's White House office. I asked David Slate, our new general counsel, to come with me and present our case to Meese and his advisers, but they didn't buy it, and the Justice Department made it clear that it would have our brief stricken if we filed it. At that point I struck a compromise with the other commissioners, making our draft brief

---

* She had done this in order to make it possible for EEOC to be put in charge of enforcement of the Equal Pay and Age Discrimination in Employment acts (both of which had previously been enforced by the Department of Labor) and given responsibility for the federal government's equal-employment opportunity and affirmative-action programs (which had been supervised by the Civil Service Commission). It was in part because of this ill-advised expansion that EEOC had grown so unmanageable.

publicly available without filing it in court. While this appeased them, it also infuriated the Justice Department, but that was their problem, not mine. They'd mishandled the situation right from the start, and I knew that I had to get along with my fellow commissioners if I wanted to accomplish anything at EEOC.

By then it was no secret that I was uncomfortable with the aggressive tack taken by some of my colleagues in the Reagan administration. One of them, as it happened, was also a black man, Clarence Pendleton of the U.S. Civil Rights Commission. I liked Penny personally and agreed with him on most civil-rights matters, but he had an unfortunate habit of making needlessly belligerent public statements on race. We were often lumped together by the press as "the two Clarences" and "the civil-rights twins," dismissed and damned as though we were interchangeable, even though Penny's public pronouncements were far more confrontational than mine. I preferred to keep a lower profile whenever I could, and I also thought that the administration's in-your-face approach to race relations was counterproductive. As a result certain of my colleagues concluded that I wasn't a team player, and I was asked to meet with several members of the administration in the White House Mess. This wasn't an uncommon request, but word got back to me that I was being "taken to the woodshed" to be taught a lesson, and one of the participants in the meeting started by telling me certain things I had to do at EEOC. I stopped him cold. "As far as I'm concerned," I said, "there are only two things I *have* to do: stay black and die." If I failed at EEOC, I added, it would be because of my own decisions, not because I was carrying water for some White House staffer.

My main quarrel with the Reagan administration was that I thought it needed a *positive* civil-rights agenda, instead of merely railing against quotas and affirmative action. This was my top priority at EEOC: to do what I could to make things better for ordinary people. It explained why I poured so much of my energy into straightening out the agency's managerial woes, and why I regarded the General Motors settlement as a prime example of what I thought that EEOC ought to be doing. But I found it impossible to get the administration to pay sufficient attention to such matters. Too many of the president's political appointees seemed more interested in playing to the conservative bleachers—and I'd come to realize, as I told a reporter, that "conservatives don't exactly break their necks to tell blacks that they're welcome." Was it because they were prejudiced? Perhaps some of them were, but the real reason, I suspected, was that blacks didn't vote for Republicans, nor would Democrats work with President Reagan on civil-rights issues. As a result there was little interest within the administration in helping a constituency that wouldn't do anything in return to help the president. My suspicions were confirmed when I offered my assistance to President Reagan's reelection campaign, only to be met with near-total indifference. One political consultant was honest enough to tell me straight out that since the president's reelection strategy didn't include the black vote, there was no role for me.

I continued to hear from Anita Hill throughout this period, almost always when she wanted something. She called the office fairly regularly, usually speaking with Diane, and she also called me at my home on occasion until I changed my phone number.* Only one of our conversations amounted to anything important, however. In 1985 Oral Roberts University moved its law school to Regent University, a Christian school in Virginia, and Dean Kothe, who stayed behind in Oklahoma, called me repeatedly after the move. I barely knew him and had no idea what he wanted. Anita called before I had a chance to get back to him and asked me to speak with him as a personal favor to her. Dean Kothe had been

* I didn't change it on her account, though—I'd been receiving late-night calls that disturbed my sleep.

good to her at Oral Roberts, she explained, and had helped her find another job at the University of Oklahoma Law School. Now he was looking for something new to do, preferably in Washington. I didn't see how I could help, but I agreed to set up a meeting with him.

Much to my surprise, Dean Kothe and I became fast friends. I learned that he was a highly successful, independently wealthy lawyer who'd set up the law school at Oral Roberts University after retiring from the practice of law. I found him to be a remarkable man—professionally, intellectually, and spiritually—and realized instantly that I could benefit from his wisdom and experience, so I brought him to EEOC in 1985 as counsel to the chairman. He worked tirelessly to understand the agency and its mission, and after he found out that my efforts to improve its operations were all but unknown to the public at large—as well as to the administration I served—he took it upon himself to set the record straight, urging me to stay on for a second term as chairman and lobbying the Reagan White House on my behalf with all the zeal of a missionary.

In May 1986, once it had become clear that I would be renominated for a second term, Dean Kothe decided to see what he could do to rally support for my confirmation in the business community, which had paid little attention to me or my work. He started by setting up a luncheon meeting with a dozen business lobbyists at the University Club. It was pouring down rain when I arrived, and only a few people showed up. One of them was Virginia Bess Lamp, a labor-relations lobbyist for the U.S. Chamber of Commerce. Virginia and I had met for the first time the previous month at a conference on affirmative action in New York City that was sponsored by the Anti-Defamation League. Midge Decter, Norman Podhoretz's wife, had introduced us, and we shared a cab to the airport after the meeting. Virginia had asked me how I coped with controversy, and I pulled out of my wallet a prayer attributed to St. Francis of Assisi that I recited daily for sustenance and guidance:

> *Keep a clear eye toward life's end. Do not forget your purpose and destiny as God's creature. What you are in His sight is what you are and nothing more. Do not let worldly cares and anxieties or the pressures of office blot out the divine life within you or the voice of God's spirit guiding you in your great task of leading humanity to wholeness. If you open yourself to God and his plan printed deeply in your heart, God will open Himself to you.*

Virginia had offered to set me up with one of the black women with whom she worked at the U.S. Chamber of Commerce. I thanked her but explained that I wasn't interested in dating. As the cab pulled up to the airport entrance, I suggested that we might "do lunch" someday, but I'd never followed up on it. Now here she was, soaked to the skin, having gone to the trouble of walking to the University Club in a torrential downpour to attend a luncheon meeting in my honor. I was mortified. As soon as I got back to my office, I told Diane to set up a lunch as soon as possible, and we met on May 29 at Hunan Rose, a Chinese restaurant on K Street. It was a pleasant, professional conversation, but nothing more. Virginia was then involved in a long-distance relationship, and I'd meant what I said when I told her I was no longer interested in dating; I'd been in two serious relationships since leaving Kathy, and neither one had worked out. I didn't want another breakup, much less a second divorce. Besides, Virginia was white, and I had no inclination to date outside my race. I had more than enough problems without adding that one to the list. For all these reasons, I felt sure that if she and I were to develop any kind of relationship, it would be comfortably platonic.

A few weeks later, the two of us went to an early-afternoon movie, *Short Circuit*. I found it hilarious, though Virginia seemed more amused by my laughter than the movie. The daylight ending to our rendezvous assured me that she would become a friend, nothing more. Not long afterward, though, we went to Baltimore's Inner Harbor and spent an afternoon talking about life, politics, and religion—and the fact that people were staring at us. I'd grown up in the Deep South, where merely being seen with a white woman was enough to get a black man lynched, but I was fascinated by Virginia, an old-fashioned idealist whom Washington's cynics had not yet managed to spoil. (They never would.) She'd only just turned twenty-nine, a single woman in a tough city, and still thought it was possible to make the world a better place for everyone. She believed in freedom and free enterprise. She wanted good to prevail over evil. She wanted America to live up to its principles, not parcel out benefits based on self-interest.

The more we talked, the clearer it became to me that even though we were different in many ways, we wanted the same things out of life. Far more important, we both wanted to be loved—by each other. At first it scared me to think that I might be seriously interested in her, but I shook it off. In my darkest hours, I'd prayed for love and happiness. Now that it seemed possible my prayers had been answered, how could I let my own fears, or the bigotry of others, stand in the way? I knew this was no fetish-laden intrigue with a woman of another race, but a gift from God. As for Virginia, she was willing to fight anybody, including friends and family, who objected to our love. Fortunately, her family never questioned her judgment, and ever since that day in Baltimore we've been joined as one, outwardly by our hands, inwardly by our hearts.

As MY FIRST term as EEOC chairman drew to a close, my friends asked why I wanted to stay on. Even Senator Danforth found it impossible to understand my tenacity. I met with him in the spring of 1986 to ask for his support in what I expected to be a very nasty confirmation battle. He agreed to help, but then he asked, with a pained look on his face, "Why are you doing this, Clarence? It isn't a good career move. Nobody likes you."

More than once I'd asked myself the same thing. I had every reason to leave EEOC, not least because the steady escalation of the ugly political battles over the president's civil-rights record had made my job an endless struggle with daily indignities. Back in 1984 I'd told Juan Williams exactly how I felt about the refusal of civil-rights leaders to treat President Reagan other than contemptuously. All they did, I said, was "bitch, bitch, bitch, moan and moan, whine and whine. That doesn't help anything. . . . You don't call the judge reviewing your case a jackass; you don't call the banker reviewing your loan application a fool. But that's exactly what black leaders have done with this administration. They've called the president everything but a child of God." When my words appeared in the *Post*, Vernon Jordan, the former head of the National Urban League, called me up and read me the riot act. He said that he would have used me for "cannon fodder" if he'd still been running the Urban League, adding that I'd need a "constituency" if I wanted to do as well as he was doing after I left EEOC. His warning was polite, firm, and clear, and I replied with equal firmness: I told him as diplomatically as I could that I wasn't looking for a constituency, nor did I care to use the woes of my people to advance my own career.

That was one of the few times I spoke out in public on this subject, but my silence did me no good. Not only did I continue to find myself on the receiving end of ad hominem attacks, but some of them came from liberal whites who would never have dared to say anything similar about a black man who shared their political views. After I wrote a letter to the editor of *Playboy*, taking issue with a 1986 article by Hodding Carter called "Reagan and the Revival of Racism," Carter responded as follows: "As a Southerner, Mr.

Thomas is surely familiar with those 'chicken-eating preachers' who gladly parroted the segregationists' line in exchange for a few crumbs from the white man's table. He's one of the few left in captivity." Not a single civil-rights leader objected to this nakedly racist language. For daring to reject the ideological orthodoxy that was prescribed for blacks by liberal whites, I was branded a traitor to my race—as if anyone, least of all a white journalist, had the right to tell me what beliefs a black man was permitted to hold. If I dared to step out of line, if I refused to be another invisible man, then I wasn't *really* black, I was an Uncle Tom doing Massa's bidding. That wasn't politics, it was hate.

The only good thing about these attacks was that they encouraged me to return to the faith that had sustained me in my youth. My troubles at EEOC had already driven me to my knees, and from there to daily prayer and meditation. Each morning I stopped at a Catholic church on my way to work and asked God for "the wisdom to know what is right and the courage to do it." It was one more giant step toward home. My closest friends were aware of my struggles, and in the end they helped guide me back to the place where I belonged. By running away from God, I had thrown away the most important part of my grandparents' legacy. Now I began to reclaim it, bit by bit. It wasn't easy for me to admit that the unsophisticated, ill-educated people among whom I'd grown up understood all along what I was only just beginning to accept, but grief is a great teacher. While I still had a long, hard road to travel before I would be fully ready to re-embrace my lost faith, the pilgrimage had begun in earnest.

I finally started to bridge the gap that separated me from the rest of the Reagan administration with the help of Ricky Silberman, who came to EEOC in 1984 as a commissioner and later served as vice-chairman. Ricky and I had met briefly at the Fairmont Conference in 1980. At the time, she was living in San Francisco, but since then she'd moved to Washington, where she worked in the Senate and, later, the Federal Communications Commission. Ricky was loyal and indispensable, and always kept the agency's best interests (and mine) firmly in mind. If I couldn't work well with someone, she did it for me, calming me down whenever I was too combative. She had an uncanny knack for smoothing ruffled feathers and seeing the positive in the ostensibly negative. It was in large part because of our friendship and her political savvy that my once-chilly relationship with the administration became closer, mutually respectful, and in the end genuinely warm.

The arrival of Bill Webb in the fall of 1982 had been another early improvement. A Vietnam veteran and former prosecutor, Bill was both independent and collegial, and he brought a sorely needed intellectual heft to the commission. Unlike some previous commissioners, neither Bill nor Ricky represented constituencies: they came to EEOC purely out of commitment and conducted themselves accordingly. So did Fred Alvarez, a graduate of Harvard Law School who began his career at the National Labor Relations Board, then worked at a large San Francisco law firm. Fred was a Democrat, but he prided himself on being a "gun-toting" moderate. I never did know what he meant by that, but from the start I knew that I could count on him to do the right thing.

The best thing Fred did for me, though, was to bring Pamela Talkin to EEOC, where she eventually became my chief of staff. Pam had previously worked with Fred in the NLRB's San Francisco office, rising from an entry-level position to become one of the most senior managers there. I was impressed by her skills and thought she'd bring a fresh perspective to EEOC management. Pam was the only person to whom I entrusted the full responsibility of running all of EEOC's operations. Each of her predecessors (Chris Roggerson, Allyson Duncan, and Jeff Zuckerman) had been outstanding, but I'd limited the scope of their authority to policy issues or enforcement matters. A self-starter and a self-described liberal, Pam agreed with EEOC's mission, understood where I wanted to take the agency, and was both totally loyal and a quick study. Some members of the Reagan administration objected strongly to hiring a liberal career employee for a position that normally would have gone to a political appointee, but I needed a proven manager to help me run EEOC, and she fit the bill. It was Pam, for instance, who helped lead the way to updating the agency's outmoded data-storage system without spending any additional money—Congress wouldn't give us one extra dime. Throughout three challenging years together, we never had a disagreement.

I'd previously been trying without success to improve EEOC's policy-making process. Now, with the support of Ricky, Bill, and Fred, the quality of our work improved dramatically. They didn't always agree with me, but we shared a collective commitment to professionalism that made it possible for us to work well together, thus freeing me to spend more time straightening out the agency's operational affairs. When I arrived at EEOC, the agency spent little time investigating cases of alleged employment discrimination. The focus was on summary settlements, and nominal relief for those who filed charges. While this "rapid charge" approach led to a large number of settlements, few cases were actually investigated. How could we claim to be enforcing the laws when we made no effort to investigate most of our cases and determine whether discrimination had actually occurred? Our enforcement program did little more than generate impressive statistics and meet the agency's numerical goals. That wasn't justice—it was pushing paper through a meaningless administrative gristmill.

With my fellow commissioners taking the lead, EEOC adopted a series of policies intended to bolster our enforcement efforts and maximize the relief obtained for those who were harmed by discrimination. It was a never-ending battle, for the mind-set that cases should be closed by any means necessary had become deeply entrenched in the agency. Every time one shortcut was blocked off, another opened up. But in time the culture of expediency gave way. Fully investigated charges had been the exception when I arrived at EEOC in 1982; they were the rule when I departed eight years later. Toughening EEOC's approach to enforcement, improving its management, and automating its data processing were our main priorities at EEOC—and our biggest successes.

To make these things happen took brutally exhausting work, though, and after a few years I came to feel that political life in Washington, whatever its other virtues, wasn't intellectually rich enough to suit me. I wanted more out of life than an endless string of twelve-hour days. In time I found it, with the unwitting help of my son. Jamal and I bought a small color television and our first VCR in late 1985, and within a few weeks I'd become fascinated by TV documentaries about World War II. How, I asked myself, had Adolf Hitler gained control of one of the most culturally advanced countries in the world? Why had so many of the world's leaders, both in America and Europe, been so slow to confront him? In an effort to answer these questions, I started watching programs about Winston Churchill, the one person who had understood and opposed Hitler early on. This led me to start reading about him, and the more I read, the more impressed I was by his prescience—and his personality. My interest in Churchill kindled a love of reading for its own sake that I'd failed to acquire in college. Before long I was gobbling up such fat tomes as Paul Johnson's *Modern Times* and *A History of Christianity*, after which I branched out to Lincoln biographies. For lighter fare I treated myself to the western novels of Louis L'Amour, and I also reread *The Fountainhead* and *Atlas Shrugged*, whose scathing criticisms of the dangers of centralized government impressed me even more after working in Washington.

I now saw that I was spending far too much time immersed in

purely practical matters—goals, timetables, quotas, budgets, personnel, accounting, computer databases—just as I'd become bored with the tired litany of overly familiar issues that had come to dominate political discussions in Washington. I was restless and hungry for intellectual challenges, and Pam Talkin gave me the opportunity to seek them out by assuming many of the day-to-day responsibilities for running EEOC after I was reconfirmed in 1986. Instead of rehashing the usual partisan debates over quotas and civil rights, I could now turn my attention to more substantive questions. Among other things, I led my staffers (especially Ken Masugi and John Marini) in discussions of the natural-law philosophy with which the Declaration of Independence, America's first founding document, is permeated. "All men are created equal," Thomas Jefferson had written in 1776. "They are endowed by their Creator with certain unalienable Rights." That's natural law in a nutshell: if all men are created equal, then no man can own another man, and we can only be governed by our consent. How, then, could a country founded on those principles have permitted slavery and segregation to exist? The answer was that it couldn't—not without being untrue to its own ideals. We debated at length the implications of natural-law thinking, and speculated on how it might apply to contemporary political discussions. These arguments stimulated my mind in a way that no discussion of current events could possibly hope to equal.

From the beginning of my time at EEOC, I'd kept a busy speaking schedule, traveling almost invariably with my confidential assistant, Armstrong Williams. I frequently spoke to internal EEOC groups as well as organizations involved in our work in Washington and around the country; I also made every effort to speak to student groups at all levels, and gave speeches during Black History Month and on various other occasions. At first I'd used speechwriters, but as time passed I started writing more of my own speeches and heavily editing the ones that continued to be written for me, and I also found time to write op-ed pieces, book reviews, and longer articles. This, too, helped me to break free from the mind-numbing effects of the daily grind of running a government agency. I felt my intellect was reawakening after years of hibernation.

ON MAY 30, 1987, Virginia and I were married in the church of her childhood, St. Paul United Methodist Church of Omaha, Nebraska. Jamal was my best man; Jill Elliott, Virginia's niece, was her bridesmaid. It was the perfect wedding for us, simple and right. As I mingled with family and friends at the reception in the church basement, I knew that I'd found my life's companion. We'd grown very close over the year that we had known each other; by now we were virtually inseparable. I loved the idea of being married to her and looked forward to the years ahead. My gloomy moods became less frequent, and for the first time since my half-forgotten days in Jefferson City, I began to enjoy myself.

At first Virginia, Jamal, and I shared a tiny two-bedroom apartment in Alexandria that I'd found the year before. We loved the location but needed a larger place, so Virginia and I started searching for a new home. I was up to my ears in debt and doubted whether I could pass the kind of credit check necessary to obtain a mortgage, but thanks to Virginia's income, our overall financial picture had improved considerably; I'd never failed to pay Jamal's tuition, my taxes, or my student loan payments, and I hoped that might make a difference as well. Early in 1988 we looked at a small model home in Kingstowne, a new development south of Alexandria. We both loved what we saw and thought we would be able to afford a small three-bedroom house, so we signed a contract and hoped for the best. As the house neared completion, though, we ran into problems with financing, and after answering countless humiliating questions about our credit history, it seemed clear that the mortgage would not be approved. Virginia and I were so despondent that we actually visited the unfinished house late one night to pray for God's intervention, and with the help of her parents and a persistent young man at the mortgage company, our prayers were answered. In May we moved into our brand-new home. It was modest in size—1,800 square feet—but after living in an 800-square-foot apartment, it felt like a mansion. I remembered how it had felt to move from Leola's tenement apartment to Daddy's gleaming white house on East Thirty-second Street in Savannah and marveled at the exciting course that Virginia and I were charting.

Virginia, Jamal, and I spent all our free time together. For years I'd been the first to arrive at the office each morning and the last to leave at day's end, but now I wanted to stay home as much as possible, so I started putting in eight o'clock–to–six o'clock days and cut back drastically on my weekend work—none of which would have been possible had it not been for Pam Talkin. I barbecued on the deck several times each week or prepared my favorite spaghetti dish; I loved planting tulips and other bulbs in the fall, mowing the lawn, and cleaning the garage. Jamal had grown into a delightful young man, creative, cooperative, and funny, and I reveled in his company. By then he was playing on his high school football team, and I went to so many practices and games that some of his teammates started calling me "Super Dad." Unlike Daddy, who had discouraged me from taking part in team sports, I loved watching him play, and I never grew tired of waiting outside the smelly locker room after each game to pat him on the back and walk proudly to the car with him.

A month after Virginia and I celebrated our first anniversary, I turned forty. I'd never had a better birthday—and I'd never been happier. At long last, I had found peace.

8

APPROACHING THE BENCH

Not only had I found contentment in my personal life, but the *Washington Post* was taking favorable notice of the changes I was making at EEOC—an unexpected switch after so many years of being raked over the coals in the media. I was flabbergasted when the *Post*'s editorial writers, who rarely had much use for anything the Reagan administration did, praised the agency in near-glowing terms. After criticizing the Civil Rights Commission, the editorial, which ran in the *Post* two weeks before Virginia and I were married, observed that

> things are markedly different at the Equal Employment Opportunity Commission. Here, the caseload is expanding and budget requests are increasing. Under the quiet but persistent leadership of Chairman Clarence Thomas, the number of cases processed has gone from 50,935 in fiscal 1982 to 66,305 last year. In the same time period, legal actions filed went from 241 to 526 . . . legislators who care about civil rights enforcement have a special obligation to sustain an agency doing this

work and enjoying, to an unusual degree in these times, the support and encouragement of the administration.

The only thing that disappointed me was that the *Post* had used EEOC's achievements as a stick with which to beat Clarence Pendleton. Penny's refusal to hew to the conventional wisdom on racial matters meant that he was subjected to constant, vitriolic attacks by civil-rights groups. In the past he'd fought back gamely and seemed irrepressible, but he told me at our last meeting early in 1988 he was afraid that the steady stream of abuse had undermined his ability to earn a living. I'd never heard him say anything like that, and his despondency alarmed me. Soon afterward he died of a heart attack. I couldn't help but feel that his tormentors had gotten to him at last.

I flew to San Diego for Penny's funeral, returning on a red-eye flight to meet Vice President George Bush the following morning. That spring Dean Kothe had introduced me to Ed Lawson, who was a friend of the vice president, during one of my trips to Tulsa. Now that President Reagan's second term was coming to an end, Dean Kothe thought I ought to get to know the vice president, and Lawson offered to set up a meeting. Unfortunately the timing was bad for both of us: the vice president was polite and personable, but Michael Dukakis had pulled far ahead of him in the polls and I could tell that he was distracted, so our talk was very brief. It was the only time I spoke to him in private until the day he announced my nomination to the Supreme Court.

I must have made a positive impression on him, though, because Reid Detchon, a former colleague from Senator Danforth's office who was now writing speeches for Vice President Bush, called a month later to ask for my assistance. The vice president would be speaking at the NAACP's national convention, and Reid wanted to know if I'd help with the speech. Some of the vice president's aides feared that he might be booed by so partisan an audience, but as it turned out, the speech was received warmly. I felt gratified that someone in the Reagan administration had seen fit to take my ideas seriously—and had gone to the trouble to consult me before the fact. Too often I found out what the administration was up to by reading about it in the *Post*. After the election I worked with Tom Gibson and H. P. Goldfield to draft a memo to the new president-elect suggesting that he take a more positive approach on racial issues. We also recommended that he consider appointing blacks to positions of responsibility other than the race-related ones they'd traditionally held. The vice president later sent me a personal note saying that he would circulate the memo to his staff.

In spite of these developments, I thought it was time to look for a new job, since I had been in Washington for close to a decade. As long as I was still at EEOC, though, I wanted to continue pushing for the appointments of minorities to senior positions, so in December I drew up a list of possible candidates (two of the names on it were Colin Powell and Condoleezza Rice) and met with Michael Uhlmann, a senior member of the president-elect's transition team, to discuss it over breakfast. I explained to Mike that it was important for the new president to place blacks in nontraditional positions, and he seemed to take what I said seriously. At the end of the meeting, he asked where I wanted to work after leaving EEOC. I admitted to being interested in serving as the deputy director of the Office of Management and Budget, but added that I knew it was a long shot and that there wasn't anything else I especially wanted to do in government. Mike asked if I would be interested in becoming a federal judge. "That's a job for old people," I said. "I'm forty. I can't see myself spending the rest of my life as a judge." Mike urged me not to dismiss the possibility out of hand, and we left it at that.

I called Virginia to tell her about the conversation, saying the same thing to her that I'd said to Mike: I couldn't see myself as a judge. She asked me to speak to Aubrey Robinson, the chief judge of the U.S. District Court in the District of Columbia, before ruling it out. Mike had already suggested that I talk it over with Ricky Silberman, who in turn recommended that I speak with her husband, Larry, a judge on the U.S. Court of Appeals for the District of Columbia Circuit. Both men told me to say yes. I explained to Larry that I was uncomfortable about accepting a lifetime appointment. "It's not like slavery, Clarence," he pointed out with a smile. "You can always leave if you don't like it." Judge Robinson, a blunt, straight-talking man, was even more emphatic: he told me that I was on a fast track and that I should definitely accept a judicial appointment if it were offered to me.

All this made sense, but I was genuinely reluctant to spend any more time in government, or in Washington. The trouble was that I didn't know what else to do. I'd gotten plenty of offers since going to EEOC, but none had excited me. In 1986, as my first term was winding up, I was sounded out by several headhunters, one of whom wanted to know whether I'd consider becoming president of one of baseball's major leagues. "Would I have to go to the games?" I asked. He said it was part of the deal, to which I replied that no amount of money could possibly make me sit through that many baseball games.

For now I was preoccupied with a project that was close to my heart: moving EEOC to a new home in downtown Washington. In 1988 the State Department had expressed an interest in taking over our Columbia Plaza headquarters. That was fine with me—I'd always thought the run-down building was awful and had promised our employees that I'd find a new one before I left the agency—so Andy Fishel, my Director of Financial and Resource Management Services, started looking for something suitable. My main requirements were that the new building be located in the District of Columbia and accessible to people with disabilities. I didn't want anyone at EEOC to have to use a separate entrance; I'd helped Dick Wieler navigate his motorized wheelchair onto the loading dock and through the back door of the Supreme Court of Missouri too many times to settle for anything less than full handicap access. It took a few months, but Andy found a building in downtown Washington that met all our needs, and the landlord was thrilled when we offered to sign a long-term lease. I made sure that the building committee included a disabled person. Shortly after I'd come to EEOC, a wheelchair-bound employee told me that the carpet in the corridor leading to my office was so soft that she found it hard to maneuver. We installed a more suitable carpet at once. This taught me how easy it is for those of us without disabilities to overlook small things that can make a big difference to handicapped individuals: too-soft carpets, too-narrow doorways and corridors, poorly placed bookshelves and door handles. To fix such problems rarely costs anything extra. The trick is to notice them in the first place, and by putting a disabled employee on the building committee, I ensured that they wouldn't be overlooked as the interior of the new building was being designed.

EEOC began moving to its new home shortly after President Bush announced his intention to nominate me to the bench in the summer of 1989. I was the last person to leave the old building. As I watched the 800 or so employees file out, I reflected with pride on the changes we'd made and the effects they'd had. Morale at EEOC had improved dramatically, and so had the quality of our work. Those who turned to us for help could now count on the help of a highly motivated team of staffers. Of course some employees had been unwilling to do their jobs—just as some people in the private sector don't do theirs. Unlike so many of my fellow Republicans, I'd never been one to complain indiscriminately about the alleged evils of lazy Washington bureaucrats. To the extent that such people existed at EEOC, it was my duty to get rid of them, and I did so. But it was also my duty to make sure everyone who remained was treated fairly and given a real opportunity to successfully carry out the agency's mission, and I did that, too. Moving and properly equipping the new headquarters was the last piece of the puzzle.

Mark Paoletta, a lawyer in the office of Presidential Personnel, called me up in March to ask for copies of my speeches and articles. He was vague about why he wanted them, but I saw no reason not to cooperate. It wasn't until afterward that I found out that he was reviewing the backgrounds of potential judicial nominees. A couple of weeks later, I was summoned to a breakfast meeting with Robin Ross and Murray Dickman, two of Attorney General Dick Thornburgh's top aides. We spoke in general terms about my tenure at EEOC. They asked whether I wanted to be a judge. The question took me by surprise, and I said I wasn't sure. My indecision was no pose: I really didn't know what to do.

One of them told me that there was some question as to whether I had enough legal experience to handle the job. I pointed out that I'd written briefs in and argued some forty-odd cases and also tried a number of tax cases during my years in Jefferson City. That was news to both men—they knew nothing of my work for Jack Danforth—and they asked if I'd be willing to fill out the forms necessary to set the nominating process in motion. I said I would, silently reminding myself that the Bush administration might well change its mind after taking a closer look at my record. As soon as I got back to my office, I called Virginia to tell her of my decision. Halfhearted though it was, I thought it might be the best course of action to let the FBI start investigating me and see where things went from there. "Maybe this is God's way of telling me what to do," I said. She agreed, and we decided to let the process run its course.

I was taken aback by the number of forms I had to fill out. In addition to the personal information I'd already supplied in preparation for my three previous nominations, the Justice Department asked for detailed information about aspects of my legal background to which I hadn't given any thought since leaving Missouri. I dug through dozens of old boxes and called friends from Monsanto and the attorney general's office in Jefferson City in order to track down documents that had long since gone astray. I also read two books about Judge Robert Bork's failed Supreme Court nomination, his own *The Tempting of America*, and *The People Rising*, written by two of his opponents, Michael Pertschuk and Wendy Schaetzel. I had watched the Bork hearings on TV, but I knew little of the concerted effort that had been made to block his confirmation, and was alarmed to find out that many of the people who had attacked me during my years at EEOC and the Department of Education had also taken part in the successful campaign against Judge Bork. Was I prepared to run the risk of being put through a similar ordeal—especially since I wasn't sure I wanted to be a judge in the first place?

In the end I decided that I was willing to run the gauntlet, and in June President Bush announced his intention to nominate me to the U.S. Court of Appeals for the District of Columbia Circuit. The usual rumbles of criticism started up promptly: I was soft on civil rights, opposed to affirmative action, and hadn't done an adequate job of enforcing the equal-employment opportunity laws. But most of the black civil-rights groups kept a low profile, just as they had during my three previous confirmation hearings, and the leaders with whom I met seemed ambivalent about the prospect of trying to scuttle my nomination. A few of them even chose to support me. I was pleasantly surprised, for example, when William Coleman, co-author of the NAACP's *Brown* v. *Board of Education* brief and the second black to serve as a cabinet member, asked me to pay him a visit. I knew him slightly—I'd gone to law school with his son and daughter—and admired him greatly. He'd been one of Judge Bork's most influential opponents, so I was pleased when he said he would support me, adding that Elaine Jones of the Legal Defense Fund would also give me discreet assistance. Elaine never told me anything confidential about the inner workings of the interest groups lining up against me, but she warned me to prepare for the worst, and I took her advice seriously.

Fred Abramson, one of Gil Hardy's many friends, was also tremendously helpful. He led the team of American Bar Association lawyers responsible for evaluating my qualifications, and more than once he quietly asked me to meet him on the street near EEOC's new headquarters so that he could tell me what the interest groups—in particular the hard-left Alliance for Justice—were up to. Fred believed that these groups were using the ABA to fight their own ideological battles, which he considered a perversion of the spirit of the review process. His distaste for their tactics was obvious, and I appreciated the sense of decency that led him to warn me of what lay ahead. Some of the reports that Fred showed me had been prepared by the Alliance for Justice, and I immediately spotted the hand of George Kassouf, a young white man whom I'd run into earlier that year when I met with Frankie Freeman, a noted veteran of the civil-rights movement, in her hospital room at Howard University Hospital. Freeman could not have been more cordial, but Kassouf treated me with palpable arrogance and condescension, so I wasn't surprised to see that he was now working against me.

One of the unpleasant realities of being a presidential nominee is that for the most part, it's up to you to get yourself confirmed. So many nominees are under consideration at any given time that there's only so much the administration can do to help any one of them. Besides, the mere fact that you've been nominated may well be enough to earn the president who sent your name to the Hill a fair amount of political credit—enough that he may not feel the need to see your nomination all the way through. But John Mackey, who spearheaded confirmations at the Justice Department, pushed hard for me, and Phyllis Berry, who'd left EEOC to spend more time with her son, volunteered to help. Having watched Judge Bork go down to defeat in the Senate, Phyllis was determined to mount a full-fledged campaign on my behalf, and I already had good reason to know that I could trust her judgment. And Pat McGuigan and Tom Jipping of the Free Congress Foundation were just as committed as she was.

Pat, who had been deeply involved in the effort to confirm Judge Bork, urged me to write a short third-person biography in order to prevent my critics from distorting my life story. I took his advice, though I deliberately understated some of the challenges I'd faced in my youth so as to avoid accusations of exaggeration. Pat reproduced it verbatim in one of his group's publications, and portions of the document, just as he'd predicted, subsequently appeared in the press. But other, less flattering stories soon followed. After the *Wall Street Journal* ran a mostly favorable article about me, a skeptical reporter from Atlanta flew to Savannah to check out my story. My mother somehow got the impression that he was a personal friend of mine; not only did she show him my childhood bedroom and give him a tour of the neighborhood, but she drove him out to Pinpoint and showed him around. The reporter later told me that his doubts were laid to rest that day, but his editor refused to let him say anything favorable about me in the piece that finally ran.

The first official indication that things would be out of the ordinary came when Joe Biden, the Democratic chairman of the Senate Judiciary Committee, presented me with a document request that entailed an enormous amount of backbreaking labor on my part. It was clearly the work of the Democratic Senate staffers with whom I'd crossed swords so many times over the years. Shortly after I received the request, I had lunch with Paul Gigot, the *Wall Street Jour-*

nal's Washington columnist. He showed me a copy of the request and told me that the *Journal* wanted to publish it on its editorial page in order to give its readers a better idea of what the Judiciary Committee was up to. I begged him to make clear that he hadn't gotten the document from me, and he promised to do so. A lot of reporters had given me similar assurances in the past, but Gigot, unlike them, kept his word. The *Journal* reprinted the entire document in small type, accompanied by an editorial which left no doubt that I hadn't leaked it.

Eventually the Justice Department swung into action, helping me prepare for the hearings and holding a couple of practice sessions. In the meantime, because of the outstanding efforts of Debbie Graham Kiley and the EEOC staff, I complied with the Judiciary Committee's document request and submitted more than thirty boxes of material, most of which was already part of the public record. I knew that most of these documents had been requested not to shed light on my background and qualifications but to give the liberal groups opposed to my nomination every opportunity to make whatever mischief they could. I often wondered who had appointed these posturing zealots to be watchdogs over the judiciary. Was anyone watching *them*, examining their backgrounds, agendas, and motives? Not that I could see. All eyes, it seemed, were fixed on me.

The most painful event of the summer, however, had nothing to do with my confirmation: Gil Hardy died in a scuba-diving accident off the coast of Gibraltar. Rarely has a piece of bad news hit me harder or left me lower. My friendship with Gil went all the way back to my Holy Cross days. Since then we'd gone to law school together, been in each other's weddings, roomed together briefly after our marriages dissolved, and spent countless hours talking about everything under the sun. I'd always assumed that we'd grow old together, too, but now I knew I'd have to face the future without him, and it was hard to imagine. I loved Gil, not least because of his warm, outgoing nature. I'd always been one to close my shutters to the world, but he threw his wide open, drinking in its goodness. His death put the passing discomforts of my confirmation battle in perspective. Next to the death of a beloved friend like him, everything else seemed trivial.

As the year drew to an end, my hearings were scheduled for February, and the partisan sniping unexpectedly eased off. I couldn't understand why until I paid a visit to Senator Biden just before the hearings started. He informed me that I would be confirmed for the Court of Appeals, but that I could expect things to be very different if I were to be nominated to the Supreme Court. That jolted me. It was the first time that the idea of becoming a Supreme Court justice had ever occurred to me, and I found it frightening. So far as I knew, nobody in the Bush administration had any such thing in mind, nor had I heard any talk that it might be a possibility—but clearly Senator Biden had. Now I began to suspect that my opponents had been sending me a message: *Don't you dare think about the Supreme Court. What you've just been through was only a taste of what will happen if you do.*

The staff of the Judiciary Committee asked me to meet with them to clear up any potential problems prior to the public hearings. The meeting, I was told, would be recorded for reference purposes, but the tapes were to be destroyed once a written summary had been prepared. I expected it to take an hour or so, but it lasted for nearly three hours, in the course of which a half dozen staffers asked a lengthy list of questions bearing mostly on my work at EEOC, along with other questions arising from anonymous tips that had been phoned into the FBI. Several Democratic staffers treated me with ill-concealed hostility. Despite their grilling, and the months of preparation that had led up to it, the actual hearings proved to be uneventful. Some of the senators' opening statements sounded like counts in an indictment, but the questioning proceeded fairly briskly and with a minimum of bombast (except for Howard Metzenbaum, who found bombast all but irresistible). By the end of the day, it was over and done with.

As I reflected on the long, unpleasant process that had led up to this brief public performance, I was struck by how easy it had become for sanctimonious whites to accuse a black man of not caring about civil rights. It was as ludicrous as a well-fed man lecturing a starving person about his insensitivity to world hunger. On the other hand, my friends from school and my previous jobs all stuck by me, and Senator Danforth never wavered. In a city full of politicians who worshipped at the altar of expediency, he was a model of loyalty and courage.

Above all there was Virginia. I needed her more than anyone, and she was always at my side. From start to finish, she was a pillar of love, strength, and support. No matter what the future might hold in store for us, be it good or bad, I knew she would be there for me, and that made all the difference. After the vote on my confirmation was delayed, she suggested that we take our first vacation together, a trip to Mexico. Senator Danforth called us in Cancun to let me know that I'd been confirmed by a voice vote. I felt more relieved than happy.

Leaving EEOC was far more difficult than I'd expected. The employees decorated all eight floors of the building at their own expense, brought in a veritable cornucopia of home-cooked food, and threw me a heartfelt going-away party. One staffer after another thanked me with warm smiles and hugs for all I had done and said that things would never be the same without me. It was a joyous farewell—but a teary one, too. I recalled that terrifying day in 1982 when I showed up for work and discovered that I'd inherited an agency in crisis. Now I was leaving behind a tightly knit group of friends who had worked hard to make the agency function when others had no faith in their ability to do so. They'd trusted me and believed in our joint mission; they knew I couldn't turn EEOC around without them, and saw that our critics were far more interested in scoring political points than in fixing a broken agency. We'd beaten the odds, and we knew it.

As I cleaned out my desk, I thought back to the only order President Reagan had given me: get EEOC off the front pages. That I had done—and kept it off, too. But I also recalled a conversation I'd had early in my tenure with a group of disabled employees who pleaded with me to make our old building more accessible. One of them was a young woman who'd suffered a spinal-cord injury in a household accident that left her a paraplegic. I told her I would do my best to help.

"The Democrats promised to help us and they didn't deliver," she said. "Why should we believe you?"

"All I can tell you is that I give you my word," I replied. "That is the most solemn promise I can make to you."

The woman left EEOC not long afterward, but returned in 1989 to tour our new, fully accessible building. "You kept your word," she said when I greeted her. I needed no more thanks than that for my eight years at EEOC.

EVEN BEFORE THE Senate finally got around to confirming me, I had to start thinking about setting up my chambers (as judges call their offices) at the court of appeals. Ken Starr had just left to become solicitor general, and he asked me to consider taking on two of his incoming law clerks, Chris Landau and Matt Sawchak. Judge Aubrey Robinson had also called me while my confirmation was pending to ask if I would hire Dorothy Barry, the secretary of Judge Barrington Parker, who was about to retire. I was glad to oblige them both. Dorothy started setting up shop as soon as I was confirmed, and by March 8, 1990, the day I joined the court, she'd pulled together enough furniture and office equipment for us to get to work.

Given my initial ambivalence about becoming a judge, I was surprised to find that I liked the job. Part of what made it so agreeable was that I got along so well with most of my new colleagues. During my first few weeks on the job, I made my way around the building, meeting the court of appeals and district court judges and many of the members of the staff. The U.S. marshals even gave me a key to their gym. "*What* gym?" Larry Silberman exploded when I told him what they'd done. He'd been a judge for five years without ever hearing that it existed. Larry never let me forget the special treatment I'd received.

The D.C. Circuit had a reputation for divisiveness, but nearly all of my fellow judges went out of their way to show me the ropes, especially Larry, who became my judicial mentor. He told me that as I considered each case that came before me, I should ask myself, "What is my role in this case—*as a judge*?" It was the best piece of advice I received, one that became central to my approach to judging. In the legislative and executive branches, it's acceptable (if not necessarily right) to make decisions based on your personal opinions or interests. The role of a judge, by contrast, is to interpret and apply the choices made in those branches, not to make policy choices of his own. By compressing all that into ten well-chosen words, Larry did more to give me a judicial philosophy than any of the futile academic debates about which I'd heard far too much while preparing for my confirmation hearings.

I quickly made friends with Dave Sentelle, a down-to-earth North Carolinian who'd previously been a lawyer, prosecutor, and federal district court judge. Dave hid his astounding intellect behind a thick southern drawl and a bottomless well of colloquialisms. I mentioned to him after we finished listening to a set of oral arguments that one lawyer's argument had contained an internal inconsistency. "Yep," Dave said, "he met himself coming back." I also became fond of Jim Buckley, William F. Buckley Jr.'s brother, whose character and decency were equaled only by his irrepressible work ethic. His red Toyota could usually be found at the courthouse on weekends or holidays. He was a scrupulous judge and a principled man, and I loved being able to boast that he and I were colleagues. Ruth and Doug Ginsburg (no relation) were equally brilliant and kind, though very different in their approaches to judging. Ruth and her husband, Marty, were quick to invite Virginia and me to their home to enjoy an evening of his fabulous cooking and their wonderful company.

Judge Steven Williams was a friendly, easygoing man with a quick smile and an inexhaustible appetite for obscure regulatory journals. A former law professor, he thrived on the most complex regulatory and administrative law cases. I also enjoyed a good relationship with Harry Edwards, the only other black member of the court. Harry was smart and, like most of us, opinionated. I'd been warned that he was irascible, but I never had any problem working with him; Harry didn't have an unkind bone in his body, though he was as aggressive in pushing his points as any of the other judges. For me it was a special honor to get to know Spotswood Robinson, one of the court's senior members, since he had also been one of the lawyers who brought *Brown* v. *Board of Education* to the Supreme Court. Judge Robinson, who earlier had taken senior status, a kind of semiretirement, labored tirelessly over his opinions, writing them out in longhand on a legal pad. He was notoriously slow to finish his work, but no sooner did I read one of his meticulously crafted opinions than I saw why it took him so long to draft them. My relationship with Pat Wald, alas, was cordial but not close. (Bob Wald, Pat's husband, later worked tenaciously against my Supreme Court confirmation. Gil Hardy and Anita Hill had been lawyers in his now-defunct firm,

Wald, Harkrader, and Ross.) As for Abner Mikva, he was the only member of the court with whom I felt uncomfortable. He had spent a decade in Congress before becoming a judge, and afterward he continued to think like an elected official, making policy decisions from the bench instead of interpreting the law. I didn't like his approach to judging—and soon I would have reason to suspect that he was even more of a judicial politician than I'd realized.*

The D.C. Circuit heard a wide-ranging array of administrative law, criminal, and civil cases. I thrived on my relative anonymity, as well as the fact that I was only responsible for my own work and that of my staff, which consisted of three law clerks and two secretaries, a far cry from the army of employees I'd headed up at EEOC. After eight years I'd become so accustomed to running a federal agency that I had no idea how draining it had been until the burden was lifted. I missed my old friends, but I was glad not to be the subject of controversies or the object of congressional attacks. In addition, I had near-total control of my schedule and did much of my work at home, often reading well into the evening or during the early morning hours. I almost felt guilty because I so rarely had to stay at the office late or go in on weekends. The icing on the cake was that Virginia had moved from the U.S. Chamber of Commerce to the Department of Labor, where she now worked as a political appointee in the office of congressional affairs. Her building was across the street from mine, so we commuted together; we could look directly into each other's offices, and often we waved at each other while talking on the phone. The hard years, it seemed, were over.

I had only one nagging reservation about my new job. What Senator Biden had said to me before my confirmation hearings

---

* Karen Henderson and Ray Randolph joined the court after I did, and we, too, developed warm relationships. (I especially appreciated the fact that Karen, like myself and Dave Sentelle, was a southerner.)

proved prophetic: no sooner did Justice William Brennan announce his retirement from the Supreme Court in the summer of 1990 than rumors that President Bush would nominate me to fill his seat started up in earnest. I had no idea who started them, much less why anybody took them seriously, since I'd only just arrived at the court of appeals and was still learning how to be a judge. I liked the fact that the Supreme Court was there to correct any mistakes I might make, and I didn't want the pressure of having the final say. I'd had enough of that at EEOC. It disturbed me even more when I heard that People for the American Way had already started making preparations to block my nomination. I was relieved when President Bush quickly chose David Souter, who had also just become a circuit court judge. Not only did I want to avoid going through a confirmation for a job I didn't want, but I was focused on Jamal, who was finishing his senior year in high school. Virginia and I had decided that he needed a little more time to grow up before starting college; Fork Union Military Academy, a century-old school in Virginia, had an excellent sports program, and we thought that it would be good for him to spend another year in high school, playing football and preparing himself for the rigors of college life.

Toward the end of 1990, several newspapers reported that I was on the short list for the next Supreme Court slot. I'd still heard nothing from the White House, but the rumors persisted. In June I went back to Holy Cross for my twentieth class reunion, and many of my classmates asked me whether it was true that President Bush was planning to tap me for the Supreme Court. I told them the same thing I told everybody else, which was that I didn't want the job and was sure it wouldn't be offered to me. On the way home, though, I ran into Paul Weyrich, the head of the Free Congress Foundation, in the Boston airport. Paul told me that he had just spoken to John Sununu, President Bush's chief of staff, who said that I would be the next nominee to the court. I felt sick.

None of my friends seemed to understand my reluctance to be nominated, perhaps because none of them knew what it feels like to be caught in the eye of a political storm. I did, though, and I'd also seen what had happened to Robert Bork and Doug Ginsburg. Even though they were outstanding legal scholars, both men had been put through the wringer—and both of them were white. By then I'd shed the last of my illusions about white liberals: I knew that their broad-mindedness stopped well short of tolerating blacks who disagreed with them. What might they try against me this time? I didn't know, but after surviving four confirmations in ten years, I was sure I didn't want to expose myself yet again to their wrath.

Later that month I attended a law-and-literature conference at Princeton University. One of the speakers was James McPherson, the author of *Battle Cry of Freedom*, a one-volume history of the Civil War that had sparked my interest in that great conflict and become one of my favorite books. I couldn't pass up the opportunity to hear McPherson speak, and I'd just bought my dream car, a Corvette ZR–1, so I decided to drive to Princeton to take part in the conference. Virginia asked me to come back the next morning to take part in the Race for the Cure, an annual event that raises funds for the fight against breast cancer. I returned just in time for the race, at which I ran into C. Boyden Gray, President Bush's counsel. I didn't know Boyden well, but we struck up a conversation and ended up walking the course together, chatting about my new job on the court of appeals. It wasn't a top-secret meeting, just a chance encounter between two middle-aged men who were no longer in good enough shape to run the distance and found that they enjoyed each other's company. I had no idea how important a role Boyden was about to play in my life.

A COUPLE OF weeks later, on the last day of the Supreme Court's term, I got a call from Chris Landau. Because it had taken so long for the Senate to vote on my nomination, Chris's clerkship had been shorter than usual, and after only three months with me, he'd gone on to clerk for Justice Antonin Scalia at the Supreme Court. He wanted to talk, and I agreed to meet him for a late lunch. He seemed tired and upset. "At least no one retired this term," he said. The news cheered me up: Virginia and I had planned a road trip to Quebec and were looking forward to a peaceful summer. But Chris had spoken too soon. When I got back to the office, Greg Katsas, one of my law clerks, met me at the elevator. "Justice Marshall just announced his retirement!" he cried out excitedly.

I had an appointment with Carolyn Wright, a black state court judge from Texas, who wanted to talk about becoming a federal judge, but I found it impossible to pay attention to her. All I could think of was the news I'd just heard: Thurgood Marshall, the first black Supreme Court justice, was stepping down after twenty-four years on the bench. My distracted conversation with Judge Wright was cut short by a call from Boyden Gray. "Are you ready for another walk around the park?" he asked. A meeting, he told me, had been scheduled at the Justice Department, and Mark Paoletta (who was now working for Boyden) would be calling within the hour to make arrangements to get me there as quietly as possible. I called Virginia to fill her in, then met Mark across the street from the courthouse. He was driving his well-used Jeep Cherokee. I realized as I got into Mark's car that I'd forgotten to tell anyone on my staff where I was going. I felt as though I was losing control of my life.

Within a few minutes, I was seated at a conference table in the Justice Department's situation room, which is soundproofed so thickly that your words seem to die as soon as you say them out loud. Boyden, Attorney General Richard Thornburgh, and a handful of senior Justice Department officials had gathered around the table. They asked me a few seemingly random questions. Who was my favorite Supreme Court member? "Justice Scalia," I said, though

I'd never really thought that much about it. Would anybody be bothered by my interracial marriage? "Only bigots and liberals!" I replied glibly. The meeting was over almost before it started. I went back to the courthouse, picked up Virginia, and drove home. On the way home, she asked me what I was going to do. Once again I had no answer: I knew I didn't want the job, but I also knew I'd need to come up with a better answer than that if I was going to say no to the president of the United States, and so far I didn't have one. Later that evening Mark Paoletta called to make arrangements for me to visit the White House the next morning. Virginia, he said, was to take me to a parking lot in Alexandria where a car from the White House would pick me up. I didn't sleep very well that night.

On Friday Virginia dropped me off as instructed, and a half hour later I was walking through a tunnel that led from the Treasury Department building to the basement of the White House. I'd been to the White House many times, but never this way. I was escorted to a windowless office and left by myself for a few hours. It seemed like an eternity. As I waited, I tried to think of a way to convince President Bush to choose somebody else. The obvious reasons were my relative youth and inexperience—I'd just turned forty-three the week before and had been on the Court of Appeals for only fifteen months—but I knew these were mere excuses. Neither then nor at any other time did it occur to me that I could not do the work of a Supreme Court justice. I'd spent my whole life coping with one challenge after another, and I knew I could handle this one as well, the same way I'd learned Latin, passed the Missouri bar exam, briefed and argued numerous cases, and straightened out EEOC. The problem was that I still didn't know whether I wanted to spend the rest of my life as a judge, and I was sure that I didn't want to run the confirmation gauntlet again.

Finally Mark came to tell me that no decision had been made, adding that he was fairly sure that if I were going to be appointed, the decision would be made later that day. I went to my chambers, clinging to the hope that Judge Emilio Garza of the Fifth Circuit Court of Appeals, the other leading candidate, would be appointed instead. *Better him than me,* I thought. I felt as though I'd dodged a bullet when I woke up the next morning without having heard anything more. "Nothing in life is so exhilarating as to be shot at without result," Winston Churchill said. I recalled his words as I looked out the window at the cloudless sky: it was Saturday, and the White House hadn't called me back. I was off the hook. Virginia and I celebrated my narrow escape by jumping in my Corvette and driving to Annapolis for brunch. I wanted to click my heels and dance in the street.

I drove to my chambers on Sunday to catch up on the work that had piled up in my absence. As I sat at my desk reading a brief, Greg Katsas ran into my chambers, looking every bit as excited as he'd been when he'd told me of Justice Marshall's retirement two days earlier. "Kennebunkport is on the phone," he blurted out. *Not again,* I thought.

Seconds later I was speaking to President Bush. "Judge, we're still thinking about this Supreme Court thing," he said. "Could you come up to Kennebunkport tomorrow to have lunch with me and talk about it?"

"Yes, Mr. President," I heard my voice saying reflexively. It felt as though someone else were doing the talking for me.

The president explained that someone would be calling me in a few minutes to make the necessary travel arrangements, then hung up. I had just enough time to phone Virginia before Michael Luttig, the assistant attorney general for the Office of Legal Counsel at the Justice Department, called to give me my new instructions. I was to meet him at seven a.m. Monday morning in the parking lot of a shopping center near his suburban home. In order to keep the media off my trail, Virginia was told to stay behind. Events were overtaking me, and once again I had the impression that somebody else had seized control of my life.

Virginia and I got up early the next morning, July 1, 1991. I still thought I wouldn't be nominated, but she insisted that I prepare a statement just in case. She handed me a legal pad and I wrote down a few sentences. She read them closely, suggesting that I add three words, "Only in America," to the next-to-last line. I put the statement in my pocket, hoping I wouldn't need it, and we drove to the parking lot where I was to meet Mike Luttig, talking and praying along the way. Mike showed up at seven on the dot. I kissed Virginia good-bye, went to Andrews Air Force Base, and boarded a government plane bound for Kennebunkport, accompanied by Boyden Gray, Dick Thornburgh, and John Sununu. I hadn't expected any of them to be coming along, and midway through the flight I started wondering why they were there. Surely I wasn't the only person going to Kennebunkport to be interviewed. Who was escorting the other candidates?

The plane landed at Pease Air Force Base in New Hampshire, where I got my first look at the new Air Force One. It was an awe-inspiring sight, gleaming in the morning sun. I remembered my very first day on Capitol Hill, and the same thought flashed through my mind that had come to me a dozen years earlier: *What a great country*. A Secret Service detail picked us up in a black SUV with heavily tinted windows, and we drove briskly to Kennebunkport. I sat in silence, feeling trapped and bewildered. Everything was happening too fast. One of the agents told me as we approached the presidential compound that we would drive through the service entrance in order to avoid the press. He handed me a folded newspaper and asked me to hold it between my face and the window, just like a white-collar criminal who didn't want his picture to be printed in the morning papers.

As we drove through the gate, I looked with surprise at the modest wooden buildings. I'd gotten the impression from the news stories I'd read about Kennebunkport that it was a lavish private resort, but it was obviously a comfortable, unassuming retreat meant for family and friends. We parked and walked along the rear of the compound to the main house that sat at the ocean's edge. I saw the president and several cabinet members and senior White House staffers sitting on the deck. Were they talking about me? Mrs. Bush came up to me with a broad smile on her face. "Congratulations," she said happily. I was stunned, and my own face must have told her what I was feeling. "I guess I let the cat out of the bag," she said.

President Bush rose and said hello. He was gracious, informal, and down to earth. "Let's have that talk," he said, taking me into the house through the kitchen. He introduced me to the chef as a White House photographer unobtrusively snapped pictures of the three of us. To my surprise, the president led me to the sitting area of the brightly decorated master bedroom. He asked me to take off my jacket, sit down, and make myself comfortable. After a few preliminary pleasantries, he said he wanted to discuss the Supreme Court vacancy with me.

"If you are appointed to the Court, could you call them as you see them?" he asked. I said that was the only way I knew how to do my job, and that it was what I'd been doing throughout my adult life.

"Can you and your family make it through the confirmation process?"

"I've been confirmed four times in the past ten years. I think we can manage once more."

"Judge, if you go on the Court, I will never publicly criticize *any* of your decisions," the president said. He repeated his words, looking straight at me. Then he said, "At two o'clock, I will announce that I will appoint you to the Supreme Court. Now let's go have lunch." I had trouble getting out of my chair. I called Virginia before sitting down to eat with President and Mrs. Bush and a tableful of cabinet members. I needed to hear her voice, to know that she was there and would always be there, no matter what. I told her that I loved her, and she said she loved me, too. I saw as I hung up that my hands were trembling.

After lunch we all walked along the driveway to the small building where the president kept his office. He'd previously scheduled a press conference in front of the building, so we entered from the rear. My presence in Kennebunkport was still a secret, and he wanted to keep it that way until the very last moment. He seemed to revel in outwitting the reporters. Mrs. Bush gave me comforting, maternal looks from time to time. At two o'clock we walked through the front door together. The press corps gasped. "It's Thomas," someone blurted out. The clicking of cameras sounded like summer rain falling on the tin roof of our hand-built house in Liberty County, the individual drops blurring together in a steady pitter-patter.

As the President introduced me to America, I thought of my wife, my grandparents, and all the other people who had helped me along the way, especially the nuns of St. Benedict the Moor and St. Pius X. Then my thoughts drifted from those who had made this day possible to those who would now try to undo it. I recalled the ants I had watched as a child on the farm, building their hills one grain of sand at a time, only to have them senselessly destroyed in an instant by a passing foot. I'd pieced my life together the same way, slowly and agonizingly. Would it, too, be kicked callously into dust?

I took the statement that Virginia had made me write out of my pocket, unfolded it, and started to read:

> As a child, I could not dare dream that I would ever see the Supreme Court, not to mention be nominated to it. Indeed, my most vivid childhood memory of a Supreme Court was the "Impeach Earl Warren" signs which lined Highway 17 near

> Savannah. I didn't quite understand who this Earl Warren fellow was, but I knew he was in some kind of trouble. I thank all of those who have helped me along the way and who helped me to this point and this moment in my life, especially my grandparents, my mother, and the nuns, all of whom were adamant that I grow up to make something of myself. I also thank my wonderful wife and my wonderful son. In my view, only in America could this have been possible. I look forward to the confirmation process and an opportunity to be of service once again to my country and to be an example to those who are where I was and to show them that, indeed, there is hope.

Once the press conference was over, President Bush called Senator Danforth, who was in Missouri that day. I got on the line and he congratulated me, sounding almost ecstatic. "Clarence, I will be *devoted* to you," he said over and over again. Twenty years before, I hadn't known whether to believe him when he assured me that there was "plenty of room at the top"; now I knew that Jack Danforth never said anything he didn't mean. But then the president called William Coleman, and I could tell by the look on his face that he didn't like what he was hearing. My heart sank. Coleman's support of my nomination to the court of appeals had been priceless. If he wasn't prepared to back me now, I was sure to be in for a long, hard ordeal.

MIKE LUTTIG AND Virginia were waiting for me at Andrews Air Force Base when I returned from Kennebunkport. I was relieved to see Virginia: I knew I would need her more than ever before in the months to come. Mike congratulated me, but his very next words filled me with foreboding. "You got this far on merit," he told me. "The rest is politics." I already knew that politics would be at the center of my life from now until the Senate voted on my nomination. But the question of merit was a different matter, one that I suspected would hang over me for the rest of my life. Because I had been tapped to succeed Justice Marshall, it was inevitable that many people would assume that race had played a role in my selection. (The second question that President Bush was asked after announcing my nomination at Kennebunkport was whether his decision had been "quota based.") I was sure that I could do the job, but there was no way I could really know what the President and his aides had been thinking when they picked me.

After I had been on the Court for about five years, I raised the topic of my nomination with Boyden Gray over lunch. He had always been candid with me, so I asked him a straight question, knowing that he would give me an equally straight answer: was I picked because I was black? Boyden replied that in fact my race had actually worked against me. The initial plan, he said, had been to have me replace Justice Brennan in order to avoid appointing me to what was widely perceived as the court's "black" seat, thus making the confirmation even more contentious. But Justice Brennan retired earlier than expected, and everyone in the White House agreed that I needed more time on the D.C. Circuit in order to pass muster as a Supreme Court nominee. Senator Biden and the rumormongers, it turned out, had known that I was under consideration long before I had any idea.

While I was at it, I asked Boyden why President Bush had described me as "the best qualified [nominee] at this time." Even I had had my doubts about so extravagant a claim. "No one ever bothered to ask what our criteria were," Boyden said. He explained that the president had been looking for someone who was not only competent at doing the job but who had also been tested in political battle and thus could be counted on not to cave in under the pressure of a confirmation battle, or to change his views after being appointed to the Court. I definitely qualified on that score: I had spent a decade in the eye of the storm at EEOC and the Department of Education, and had never compromised on matters of principle. "It also mattered that your FBI file was very clean, at least as FBI files go," he added. "Everything added up to make you the best-qualified choice."

It might have reassured me had I known these things at the time—but I didn't. All I knew was that I was about to face a battle far worse than anything I'd ever before experienced. "You know that some of my opponents are going to try to kill me," I told Virginia as we drove home from Andrews Air Force Base. Of course I didn't mean it literally, but I did feel that there was a sense in which my whole life was at stake. I'd grown up in a part of America where a black man was defenseless against the accusations of any white person—especially a woman. The fear and vulnerability that I had known then came back to haunt me now. By the time we got to the house, it was almost dark, and I was startled to see a woman with a camera waiting near our driveway. She introduced herself as a *New York Times* photographer. At first we asked her not to take our picture, but she was so polite and respectful that we finally agreed, unable to imagine that our small yard would soon be trampled by hordes of photographers and reporters, some of whom would be far less polite. We spent the rest of the evening answering congratulatory calls from family, friends, and neighbors, all of whom had seen me on TV. To a person they were surprised and happy. I appreciated their calls, but I didn't share their joy. They took it for granted that I'd be confirmed without any serious opposition, for they saw my nomination as an affirmation of the American dream: a poor black child from the segregated South had grown up to become a Supreme Court justice. Who could be against that? I didn't have the heart to tell them that some liberals saw things differently.

The next morning I showed up for work as usual. It was Judge Ruth Bader Ginsburg's custom to exercise in an empty room across the hall from my chambers, and as I sat down at my desk, she hurried into my office in her exercise clothes, beaming with joy as she congratulated me. In the days and weeks that followed, I was deluged with calls, letters, and telegrams from all over the country. I heard from long-lost friends, classmates, and relatives, as well as thousands of citizens whom I had never met. It didn't take long for me to grasp the fact that being a nominee was a full-time job, and after consulting with some of my fellow judges, I decided that it would be inappropriate to write any more opinions while waiting to be confirmed. Confirmation by the Senate, I knew, was a political process, which made me a political figure. I couldn't go back to being an impartial judge until the votes were cast, one way or the other.

Now that my judging days were temporarily behind me, I started meeting each day with Mike Luttig and other Justice Department lawyers. Much of our time together was spent clearing up various rumors about me, most of which were self-evidently absurd and a few of which were downright inexplicable. Among other things, the White House was told that I had an outstanding federal tax lien. This made no sense at all, since the IRS had certified that my taxes were current prior to each of my four previous confirmations. Apparently a reporter digging for dirt in my divorce records in Maryland had stumbled across a lien notice filed sometime in the early eighties. I spoke to a Department of Treasury spokesperson who confirmed that my taxes were paid up. He didn't know why the lien had been filed in the first place, but assured me that it had been withdrawn as soon as it came to light.

The next allegation was equally nonsensical. One of my friends from the attorney general's office in Missouri mistakenly told a reporter that I'd kept a small Confederate flag on my desk when we worked together. Even if he'd been right, I didn't see why it should have mattered—surely no sane person could think I'd been a fan of the Confederacy—but of course he was wrong. The "Confederate flag" he remembered was in fact the miniature state flag of Georgia that I'd brought back from Savannah to show Joel Wilson, after which I put it in the mug that held my pens and pencils. The media, of course, jumped all over the story, tracking down so-called experts who'd never met me and inviting them to sound off about the psychological implications of this nonevent. It took a letter from Joel to put an end to the story.

After that I was accused of being anti-Semitic because I'd praised Louis Farrakhan, the leader of the Nation of Islam, almost a decade earlier. Once again the truth was infinitely less shocking: I'd been attracted to the Black Muslim philosophy of self-reliance ever since my radical days in college, and I'd made my favorable comments about Minister Farrakhan in the early eighties, at a time when I was under the mistaken impression that he'd abandoned his anti-white, anti-Semitic rhetoric in favor of a positive self-help philosophy. Nobody who knew me thought that I was anti-Semitic. Far from it: everyone knew that I abhorred the idea of racial, ethnic, or religious stereotypes, and much of the criticism I had received over the years had arisen from my refusal to make decisions on the basis of such characteristics. (My critics, by contrast, took it for granted that all real blacks thought the same way about everything.) My former colleagues, including EEOC's vice-chairman and two chiefs of staff who were Jewish, publicly refuted the claim that I was prejudiced against Jews, but the slur continued to circulate until friends at EEOC dug up an interview I'd given to a Catholic publication in the eighties in which I disavowed my favorable comments about Farrakhan, explaining that my admiration for his self-help philosophy had been rendered irrelevant by his continued bigotry.

Soon afterward a number of friends, EEOC employees, and former colleagues told me that they had been approached by a parade of reporters asking the same question: "Do you have any dirt on Thomas?" They all said that the same reporters lost interest as soon as it became evident that they liked and admired me. "They're gunning for you," one of them said. He wasn't telling me anything I didn't already know.

Any remaining questions about how I would be treated on Capitol Hill were answered when the Democrat-controlled Judiciary Committee scheduled the confirmation hearings for September, thus giving my critics the whole summer to sling their mud. But Mike Luttig promised to do everything he could to help me prepare for the coming ordeal, and I promised in turn to do everything I could to prepare myself. I kept Bobby Knight's saying in mind throughout the coming weeks: the important thing in life is having the will to prepare. (Mike was well aware of the problems I faced, since President Bush had nominated him in April to fill a newly created seat on the Fourth Circuit Court of Appeals.) At first we met each day to work on questionnaires and document requests—it took us nearly seventy pages to answer the questions submitted by the Judiciary Committee—and do what we could to quash the rumors that continued to surface throughout the summer.

I was simultaneously being evaluated by the American Bar Association, and early on I had lunch with the two lawyers in charge of the ABA's investigation. Judah Best struck me as a professional whom I could trust to be fair, but I was suspicious of Robert Watkins, a black attorney from a large law firm, Williams and Connolly. Although Watkins was polite enough, his manner was guarded and distant, and he later abstained from voting on my nomination on the grounds that he sat on the board of the Lawyers Committee for Civil Rights, a group that was adamantly opposed to me. Needless to say, he hadn't found it necessary to mention this conflict of interest until he'd finished investigating me.

Watkins asked whether I'd ever used illegal drugs, the same question that had ostensibly short-circuited Doug Ginsburg's confirmation in 1987. I said that I didn't recall ever having done so. It was an uncharacteristically cautious reply, but I'd been a heavy drinker in college and had often been around people who smoked marijuana and hashish. I was telling the truth: I didn't *remember* using such drugs. I'd been afraid of them. For me illegal drugs were yet another problem I didn't need to add to my already long list. On the other hand, it occurred to me that I might possibly have tried them once or twice while I was drunk, and I knew that a flat denial might put me at risk of being contradicted, so in the end, in order to put the issue to rest, I said that I had experimented with marijuana.

By then I'd started making the rounds on Capitol Hill. I paid more than sixty courtesy visits to senators, accompanied by Senator Danforth and Fred McClure, the head of the White House office of legislative affairs. A cluster of photographers and cameramen met me at each senator's doorway, but once they'd taken their pictures and asked their questions, the private meetings that followed were courteous, and some turned out to be quite agreeable. I thoroughly enjoyed my visit with Pat Moynihan, for example, whose work on the black family I had long appreciated and defended—but then, Senator Moynihan was a bona fide intellectual who took ideas seriously, unlike some of his colleagues who saw them as nothing more than blunt instruments to be used on their enemies.

Howard Metzenbaum was the other kind of senator, and I already knew how he felt about me. It would have been charitable to call him unlikable, though he went through the motions of civility during my visit. At one point he actually tried to lure me into a discussion of natural law, but I knew he was no philosopher, just another cynical politician looking for a chink in my armor, so all I did was ask him if he would consider having a human-being sandwich for lunch instead of, say, a turkey sandwich. That's Natural Law 101: all law is based on some sense of moral principles inherent in the nature of human beings, which explains why cannibalism, even without a written law to proscribe it, strikes every civilized person as naturally wrong. Any well-read college student would have gotten my point, but Senator Metzenbaum just stared at me awkwardly and changed the subject as fast as he could.

Some senators tried without success to conceal their political agendas. Howell Heflin, for instance, called me back for several follow-up meetings, but it soon became evident that his sole purpose in continuing to meet with me was to find reasons to vote against me. Bob Packwood, on the other hand, was direct: he said that he liked me, agreed with many things that I had said, and thought that I would be a fine member of the Court, but that he couldn't vote for me because his political career depended on support from the same women's groups that were opposing my nomination. Al Gore was equally candid when a friend of mine approached him, saying that he'd vote for me if he decided not to run for president. Strange as it may sound, I appreciated that kind of honesty. It took a certain amount of courage for these senators to admit their real reasons for voting against me instead of making up some transparent excuse.

Fritz Hollings was as sincere as he was funny. He told me that he was willing to vote for me, but that he had a political problem with the NAACP in his home state of South Carolina, which would have to be resolved before he could commit himself. "That's political with a small *p*," he added. I had testified before Senator Hollings on a number of occasions when I defended EEOC's budget requests in the Senate, and he'd been honest when it came to making budget decisions. I felt confident that he'd treat me the same way now.

After spending nearly a month filling out forms, visiting senators, meeting with other dignitaries, and stamping out rumors, Mike Luttig told me that the time had come to start preparing in earnest for the hearings. I felt as though I was entering the second stage of a triathlon, staggering out of the water and climbing onto my bicycle for a hundred-mile ride, knowing that I still had to run a marathon after that. Between late July and early September, Mike showed up at my house each morning, lugging a seemingly endless series of three-inch-thick binders crammed full of Supreme Court decisions, law-review articles, and chapters from treatises on constitutional law. I read everything he gave me, then discussed it in detail with him and John Harrison, one of Mike's staff lawyers. Sometimes they had no answers and said so, and on the subject of abortion, the issue with which my opponents were obsessed, they were careful to say nothing at all. They supplied me with copies of the relevant cases but made a point of not asking for my opinion or giving their own, thus allowing me to say truthfully that I hadn't been coached.

Many people were skeptical when I later claimed that I'd never discussed *Roe* v. *Wade*. The fact was that I'd never been especially interested in the subject of abortion, and hadn't even read the decision until it turned up in one of Mike's many binders. In law school I'd been a self-styled "lazy libertarian" who saw abortion as a purely personal matter. Like most Americans I had mixed emotions about it, and I wasn't comfortable telling others what to do in difficult circumstances. The closest that I ever came to talking about abortion at Yale was the course work I did on substantive due-process cases and the right to privacy, but *Roe* was handed down after I studied constitutional law, so it wasn't part of the curriculum. Of course I knew and understood the personal pain of those who had to choose between having a child or an abortion, but at the time I took the easy way out by remaining agnostic on the matter. The only time I can remember discussing abortion rights with anyone prior to my Supreme Court confirmation hearings was when Attorney General Danforth was preparing to argue an abortion case in the Supreme Court, *Planned Parenthood of Central Missouri* v. *Danforth*. He stood in the doorway of my office and tried out one of his arguments on me, explaining that the federal government had no business telling the state of Missouri how to regulate abortion. I replied that the state had no business telling women what to do with their bodies. He turned and left.*

My phone rang nonstop all summer long. Insofar as possible I ignored the media, but I never failed to return calls from Mike, who peppered me with questions about my long record of public utterances. At one point his office compiled a twenty-page list of potentially controversial excerpts from the speeches I had given at EEOC. "Why did you say all those things?" he asked me. In response I asked him if he disagreed with any of them. "No," he replied, "but I certainly wouldn't have said them in public!" Another time he asked me how I'd found the time to write and deliver more than a hundred printed speeches. Trying to make light of his exasperation, I told him that I would have written more if I hadn't run out of time. He wasn't amused, though, and I knew it was no laughing matter. Both of us were well aware that my opponents were sifting through those same speeches, looking for ammunition.

I got up at four a.m. most days, sat down at the kitchen table, and started reading. Whenever I needed a break, I'd take a binder of cases into our small backyard, doing my best to steer clear of the photographers and other media types who showed up unannounced at the house. Mike and John came over around ten, and we'd sit down at the table and spend the rest of the morning talking over everything I had just read. After they left I drove over to Mount Vernon to eat lunch at a picnic table next to the Potomac River. As soon as I got back home, I would deal with the most urgent of the

---

* In 1992 I joined Chief Justice William Rehnquist, Justice Byron White, and Justice Antonin Scalia in dissenting from *Planned Parenthood* v. *Casey*, the Supreme Court decision in which a majority of justices voted to reaffirm *Roe* v. *Wade*. By then I'd had ample time to study *Roe* in detail, and concluded that it was wrongly decided and should now be overruled.

messages that had piled up in my absence, then sit down at the kitchen table and start reading again.

This routine soon took its toll: I grew flabby from lack of exercise, and my stomach was in knots. I rarely went anywhere in public except for my daily drive to Mount Vernon, and Virginia took over the routine errands that we'd long enjoyed doing together. Instead of mowing the lawn, washing the car, or taking Virginia out to dinner or a movie, I hunkered down and worked without cease, flinching whenever I heard a ringing phone or a knock on the door. I prayed and meditated several times each day, but Virginia and I stopped going to church in order to spare our congregation the ordeal of being harassed by reporters. (I had been told that at least one article falsely suggested that the members of the church we were then attending engaged in such extreme religious practices as handling poisonous snakes and speaking in tongues.) By midsummer our once-cheerful home had become a joyless hermit's cell.

I'd warned the members of my family to be careful about talking to the press, and Myers and C took my advice to heart.* Jamal was safely out of reach at Fork Union Military Academy, and while the press hounded my ex-wife, publishing stories falsely claiming that I'd beaten her and that our divorce had been protracted and bitter, she told them nothing and turned them away. Not all of my relatives in Savannah were so prudent, though: several of them talked freely to the reporters who came to see them, and later regretted it. The way my sister was treated was especially contemptible. Plying her with kind words and a few hundred dollars, they did their best to use her to show that I had been a bad brother, somehow getting her to say that she'd had an abortion, which was both untrue and nobody's business but hers. Few showed any interest whatsoever in

* I'd previously visited C at his home in Philadelphia in the early eighties. It was the first time I'd seen him face-to-face since 1967. We stayed in touch for the rest of his life—he died in 2002—but never became close.

reporting on my family's real-life struggles with poverty, prejudice, and poor education. Our personal lives were of interest only to the extent that they could be used to smear me: I'd succeeded in escaping the disadvantages of my childhood in Pinpoint, but some of my relatives hadn't been so fortunate, thus proving that I'd turned my back on them in my ruthless climb to the top of the heap. Even for a journalist, that was low.

Yet the summer of 1991 had its high spots, too, one of which was the day I met with a group of family, friends, and supporters who came to Washington from Georgia in July, led by Jack Fuller, my ninety-one-year-old cousin from Liberty County. Cousin Jack, who lived about a quarter of a mile from our house on the farm, had patiently taught me how to tie fence wire and fodder and grind sugarcane. I'd spent many hours walking behind Daddy as he worked Lizzy, Cousin Jack's old plow horse. Seeing him and my other relatives gave my spirits a timely lift. Since I couldn't go back to Georgia, Georgia had come to me. It helped, too, that I heard from so many other friends that summer, some of whom I hadn't seen in decades, and on occasion I even read news stories that contained encouraging words from half-forgotten people who had passed through my life long ago. One was a young black man who had worked at the Missouri Highway Patrol gym in Jefferson City. I'd urged him to keep on reading and dreaming, and now he told a reporter that he'd since earned his Ph.D., become a college professor, and wished me the best of luck. Naturally I appreciated his support, but it meant far more to me to learn of his success, and I held on to his story like an amulet throughout my ordeal. It moved me beyond words to have been given so tangible a sign of the hope I'd talked about in Kennebunkport, a reminder that in America, dreams can still come true.

By that time I was reading the papers only when it was absolutely necessary, but Virginia made a point of passing on such stories in order to cheer me up, and I could tell from them that my friends were doing their best to get across the message that I wasn't Frankenstein's monster but a perfectly normal human being. What they didn't understand was that my opponents didn't care who I was. Even if they had wanted to know the truth about me, it would have made no sense to them, since I refused to stay in my place and play by their rules and was too complicated to fit into their simple-minded, stereotypical pigeonholes. In any case, I couldn't be defeated without first being caricatured and dehumanized. They couldn't deny that I had a loyal and loving family, so they found ways to use it against me; they couldn't deny that I'd been born into rural poverty, so they cast doubt on everything I'd done since leaving home, twisting and belittling my escape from the poverty and ignorance of my young years. Above all they couldn't allow my life to be seen as the story of an ordinary person who, like most people, had worked out his problems step by unsure step. That would have been too honest—and too human.

In addition to meeting with most of the senators who would vote on my confirmation, I also met privately with the leaders of several civil-rights groups at the home of my friend Connie Newman, another of President Bush's black appointees. As was the case with so many of my friends, Connie and I had our political disagreements, but we both understood that true friendships are built on bonds of affection and character, not shared ideology. Some of these meetings proved to be unexpectedly useful. Vernon Jordan, for example, was considerate and generous with his counsel, though he made it clear that he thought little of my political beliefs. I also met with several board members of the NAACP, but that was a waste of time, since the organization announced its opposition to my nomination shortly after the meeting, apparently at the insistence of the AFL-CIO. Friends of mine who were close to both organizations gave me a copy of the union's letter to the NAACP. They explained to me that the AFL-CIO's leaders wanted the NAACP to give them cover to oppose me at the union's upcoming convention. I was no politician, but even I could understand that maneuver: by publicly washing its hands of me, the NAACP was in effect giving a green light to the various groups that opposed my nomination, tacitly assuring them that it was now all right for them to smear a black man.

I was disappointed by the decision, but hardly surprised. I knew the NAACP had undergone profound changes since the long-ago days when Daddy and the rest of our family had been proud members. Back then it was impossible to question the rightness of its brave battles against lynchings, segregation, and discrimination. Now, though, its leaders were increasingly preoccupied with more complex social, political, and ideological matters, many of which they presented inaptly as racial issues. The controversial stands they were now taking offered much ground for disagreement and little prospect for measurable progress, so there was nothing remarkable about the fact that I differed with the NAACP so often. Nor did I question its right to oppose me based on these disagreements. What saddened me was the fact that an organization whose independence had once been a byword in the Deep South had been reduced to doing the bidding of the AFL-CIO.

It also didn't surprise me when the ABA's evaluation team announced that it considered me to be "qualified" for the Supreme Court. By withholding its highest rating, "well qualified," the ABA opened the door for my liberal opponents to attack my competence, and they immediately obliged. "The country and the Court deserve better than a minimally qualified justice," Nan Aron of the Alliance for Justice told the *Wall Street Journal*. Far worse things were being whispered around Washington, including slanderous rumors that I'd used cocaine and had sex with underage girls. It didn't help that a friend had thoughtlessly told a reporter I had talked about X-rated movies while I was at Yale. It was true—but so had many other young people in the seventies. Those were the days when *Deep Throat* was one of the most talked-about movies in America, so much so that it became the code name for the then-unknown informant who helped break the Watergate story. Of course it had been immature of me even to mention such films, but I *was* immature, like many other students. My friend's ill-judged comment, which she had made solely to demonstrate that I'd been a "regular guy" at Yale Law School, would later be used as a pretext for the media to conduct intrusive inquiries into my personal life.

I found it ironic that the same people who complained bitterly about the government's peering into their bedrooms had no objection to invading my own now that it suited their purposes to do so. I'd always done my best to keep my private affairs to myself, rarely discussing them with anyone but my closest friends. Of course I had my share of romantic involvements in between marriages, but there was nothing peculiar about that: I was a divorced man in his thirties. The important thing was that I had never behaved inappropriately toward any woman, and I had no intention of letting my enemies hang that age-old charge of sexual impropriety around my neck. Those who wished only to exploit my past failings, not forgive them, would get no help from me.

By summer's end my critics seemed like the low-country gnats that infested our Liberty County farm. I remembered how they'd swarmed at me from every direction as I walked through the fields in the cool hours before dawn. This wasn't so different—except that these gnats were lethal. To be sure, their attacks were wide of the mark, but that didn't matter: all they had to do was get enough people to believe them and the damage would be done. This knowledge filled me with paranoia, and I spent plenty of time talking to my friends on the phone in order to preserve my emotional balance. One person who helped me stay on an even keel was Diane Holt, my secretary at the Department of Education and EEOC, whom I'd long since come to regard as more like a sister. At one point she tried to make me laugh by telling me that Dave Kyllo, another former EEOC staffer, had asked her if Anita Hill would say anything negative about me. Diane knew that the question would strike me as ridiculous, and it gave us both a chuckle. I'd actually penciled her in as a liberal whom I could call as a witness on my behalf should it become necessary. Less than a year earlier she'd called my chambers to ask if I'd speak at the University of Oklahoma Law School, and I told my secretary to assure her that I'd be happy to oblige once I was settled in my new job. I was so busy swatting gnats that it never occurred to me for a moment that Anita might become one of my biggest disappointments and my most traitorous adversary. The joke would be on me.

As THE DATE of the hearings drew nearer, I took part in two "murder boards," in which a dozen experienced lawyers asked me tough questions in order to prepare me for the real thing. One person asked whether I had ever expressed any opinion about *Roe* v. *Wade*. I said that I had not. I was told that my answer didn't sound credible, to which I replied that no matter what it sounded like, it was true. After the second session, Mike Berman, a Democratic lobbyist who took part in the murder boards, took me aside and said, "No matter what happens, remember to let your friends defend you." Then he said it again. It was as if he knew something he couldn't tell me, which unnerved me. Why should my friends *need* to defend me? Wouldn't my testimony be sufficient to make the case for my confirmation? Or was some unforeseen horror lurking in the wings?

A few days before I faced the Judiciary Committee, Joseph Biden invited Virginia and me to tour the Caucus Room in the Russell

Senate Office Building, where the hearings would take place. I'd worked in that building as a staffer for Senator Danforth, never imagining that the day would come when I'd be the main attraction in the Caucus Room (or anyplace else, for that matter). Senator Biden was reassuring, stressing that the hearings weren't meant to be an ordeal. He said that since I'd be nervous at first, he would start the questioning with a few "softballs" that would help me relax and do my best, assuring me that he had no tricks up his sleeve.

Not surprisingly, my advisers were less sanguine, warning me to expect to be probed closely about two things, my temper and my views on natural law. I didn't recall having raised my voice during my professional career other than in laughter, so I couldn't see why I should worry about being criticized for the alleged shortness of my temper. As for natural law, I knew perfectly well that it was nothing more than a way of tricking me into talking about abortion, since many Catholic moral philosophers saw the two things as intimately related. But my interest in natural law was different, and I hoped that I could quell any anxieties resulting from it. If some senators found the subject silly or radical, I was prepared to oblige them by discussing the silliness and radicalism of the Founding Fathers who had written natural-law philosophy into the Declaration of Independence. Why shouldn't a federal judge be interested in what the Founders thought about natural law—and why shouldn't a black man be interested in the fact that the philosophical underpinnings of the Constitution had been in direct conflict with the peculiar institution of slavery, thus fueling the earliest efforts to free my forebears?

In addition I expected to be attacked as unqualified to sit on the Supreme Court. What this *really* meant, of course, was that I dared to hold views of which my opponents disapproved. Had I been a liberal, they would have overlooked my youth and comparative inexperience, not to mention the fact that I'd been admitted to Yale Law School in part because I was black. That painful truth was the soft underbelly of my career, and I knew that my opponents would thrust a dagger into it if they could figure out how to do so. As for my inexperience, I thought it was fair enough to bring it up; I'd been away from the practice of law for more than a decade before moving from EEOC to the court of appeals, on which I had sat for just over a year. But I knew there was more to it than that. Nobody—myself included—knew how I'd rule on abortion-related issues, but liberal activists knew that I disagreed with them on other matters, so they took it for granted that I'd also disagree with them on *Roe* v. *Wade*. Since most of them saw abortion rights as the single most important matter likely to come before the Supreme Court, I had to be stopped, whatever the cost. Years later a young woman who had worked for one of the many groups opposed to my nomination approached Virginia. "We didn't think of your husband as human, and I'm sorry," she said, tears streaming down her face. "We thought that anything was justified because our access to abortions and sex was at risk." The woman went on to explain that she had subsequently had a religious conversion and now felt that it was her duty to apologize to us.

I already knew from studying Judge Bork's confirmation hearings that the smear campaign mounted by his opponents was intended to make him look so bad in the public eye that southern senators could vote against him without running the risk of antagonizing their conservative supporters. My opponents hoped to do the same thing by charging that I was unqualified to sit on the Court—and by making discreet, strategically placed mentions of the fact that my wife was white. It was a subtle way of tapping into old racial prejudices, and I suspected that it might work.

I learned other lessons from the Bork debacle. One of them was that there was nothing to be gained from engaging in extended debates with the members of the Judiciary Committee. A confirmation hearing, Mike counseled me, is an ordeal to be endured, not an opportunity to engage in thoughtful public discussion. Judge Bork had made the mistake of patiently trying to explain the nuances of constitutional law to his questioners, despite the fact that most of them didn't know or care what he was talking about. The paradox was that my comparative inexperience might make it easier for me to stay out of that swamp. I wasn't a theorist of constitutional law, meaning that I couldn't have emulated Judge Bork even if I'd wanted to do so. Instead I planned to say as little as possible about such matters. To do more would be to run the risk of giving my enemies more ammunition to fire at me—which was, of course, the whole point of their questioning.

All these things were running through my mind on the morning of September 10, 1991. Senator Danforth would be escorting me to the Caucus Room, and Virginia and I went to his office early. He and his wife, Sally, were already there. True to his word, the senator had stood by me all summer long, talking to his fellow senators and campaigning vigorously for my nomination. Now he asked Virginia and me to follow him into the small private bathroom in his office. "You'll probably think I'm strange to ask you to do this," he said kiddingly, but by then I wasn't able to see the humor in much of anything. As soon as the four of us had crowded into the bathroom, he pulled out a portable tape recorder and played us a recording of "Onward, Christian Soldiers" by the Mormon Tabernacle Choir. The look on his face told me that this was no joke. Virginia and I listened intently to the hymn's long-familiar words: *Onward, Christian soldiers, marching as to war, / With the cross of Jesus going on before. / Christ, the royal Master, leads against the foe; / Forward into battle see His banners go!* Then Senator Danforth prayed that the day would go well, told me to go forth in the name of Christ, and implored me to let the Holy Ghost speak through me. I knew he was an Episcopal minister—it was no secret—but this was the first time I'd seen him in that capacity, and I felt blessed.

In the dark days to come, he would be everything I needed: friend, counselor, senator, lobbyist, and priest.*

As we left the office, I saw that a throng of supporters had lined the hall to wish me well. A few of their faces were familiar, but most were strangers to me. I asked Virginia who they were. "They're angels!" she replied happily. Their presence boosted my sagging spirits, but I was still half-frozen with fear. My stomach heaved and my legs felt like they were carved out of wood. I wondered if condemned men felt this way as they walked to the gallows. We passed through the doors to the Caucus Room and I saw that it was filled to overflowing. The fourteen members of the Judiciary Committee, seated together in a long row at the front of the room, loomed larger than life, dwarfing the small table and single chair that had been set up for me. Jamal, my mother, and my sister were seated near the front of the room. Virginia joined them, and Senator Danforth and my other advisers took seats immediately behind me. My old friend Harry Singleton stood next to a pillar, looking like a sentry. (He would stand there throughout the hearings, as though he was guarding my back.) I greeted the senators and went to my chair, drawing momentary comfort from the sight of my wife, family, and friends. Then I turned my back on them and sat down to face the unknown.

The morning was taken up by the committee members' lengthy opening statements, more than a few of which were so blatantly hostile as to border on the comical. I endured them as I had endured the slights and slurs I had heard all my life, but I found it of-

---

* I should also say a word about Sally Danforth. Neither Virginia nor I knew her very well prior to the hearings—she'd been busy raising her young children throughout the years I worked for her husband—but she, too, stood by us, bringing food, answering phone calls, fending off reporters, reassuring the anxious members of my family, and doing anything else that needed doing. No matter how bad things got, she was always there to offer us a helping hand and a sympathetic ear.

fensive that this particular group of people was talking about me in such terms. What gave these rich white men the right to question my commitment to racial justice? Was there no limit to their shamelessness? Not until three in the afternoon was I allowed to speak. Vernon Jordan had suggested that I express my appreciation of the civil-rights pioneers who had done so much to change America for the better, but I needed no prodding from him to do so, just as nobody had to remind me to tell the senators where I came from and what I believed: "A judge must get the decision right because, when all is said and done, the little guy, the average person, the people of Pinpoint, the real people of America will be affected not only by what we as judges do, but by the way we do our jobs."

Senator Biden was the first questioner. Instead of the softball questions he'd promised to ask, he threw a beanball straight at my head, quoting from a speech that I'd given four years earlier at the Pacific Legal Foundation and challenging me to defend what I'd said: "'I find attractive the arguments of scholars such as Stephen Macedo, who defend an activist Supreme Court that would . . . strike down laws restricting property rights.'" That caught me off guard, and I had no recollection of making so atypical a statement, which shook me up even more. "Now, it would seem to me what you were talking about," Senator Biden went on to say, "is you find attractive the fact that they are activists and they would like to strike down existing laws that impact on restricting the use of property rights, because you know, that is what they write about."

Since I didn't remember making the statement in the first place, I didn't know how to respond to it. All I could say in reply was that "it has been quite some time since I have read Professor Macedo . . . But I don't believe that in my writings I have indicated that we should have an activist Supreme Court or that we should have any form of activism on the Supreme Court." It was, I knew, a weak answer. Fortunately, though, the young lawyers who had helped prepare me for the hearings had loaded all of my speeches into a computer, and at the first break in the proceedings they looked this one up. The senator, they found, had wrenched my words out of context. I looked at the text of my speech and saw that the passage he'd read out loud had been immediately followed by two other sentences: "But the libertarian argument overlooks the place of the Supreme Court in a scheme of separation of powers. One does not strengthen self-government and the rule of law by having the nondemocratic branch of the government make policy." The point I'd been making was the opposite of the one that Senator Biden claimed I had made.

Throughout my life I've often found truth embedded in the lyrics of my favorite records. At Yale, for example, I'd listened often to "Smiling Faces Sometimes," a song by the Undisputed Truth that warns of the dangers of trusting the hypocrites who "pretend to be your friend" while secretly planning to do you wrong. Now I knew I'd met one of them: Senator Biden's smooth, insincere promises that he would treat me fairly were nothing but talk. Instead of relaxing, I'd have to keep my guard up.

Democratic senators spent the next few days pummeling me with loaded questions. Many were halfhearted retreads from my last confirmation hearing, but no sooner was the subject of abortion broached than the room heated up considerably. Patrick Leahy quizzed me aggressively about whether I'd discussed *Roe* with anyone. "Have you ever had discussion of *Roe v. Wade* other than in this room?" he asked sarcastically. "In the seventeen or eighteen years it's been there?" I explained that while I might have mentioned it in passing, it wasn't a case about which I'd thought deeply or whose merits I'd had occasion to consider. It was obvious that the senator didn't believe me. Apparently few people in Washington thought it was possible to live and breathe without debating *Roe* and forming a considered opinion on it. But that was what I'd done, so I didn't give in to Senator Leahy's bullying. All I could do was keep on telling the truth, just as I had told it to my questioners at the murder boards.

Each day I left the Caucus Room tired, tormented, and anxious, and each day Virginia and I bathed ourselves in God's unwavering love. I knew that my team was doing all they could for me, but the long months of preparation had worn me down to a shadow of myself, and I knew that no human hand could sustain me in my time of trial. After years of rejecting God, I'd slowly eased into a state of quiet ambivalence toward Him, but that wasn't good enough anymore: I had to go the whole way. I recalled one of Daddy's sayings, "Hard times make monkey eat cayenne pepper." Now, with Virginia at my side, I ate the pepper of faith—and found it sweet.

Psalm 57 showed me the way:

I will take refuge in the shadow of your wings
until the disaster has passed. . . .
I am in the midst of lions;
I lie among ravenous beasts—
men whose teeth are spears and arrows,
whose tongues are sharp swords.
They spread a net for my feet—
I was bowed down in distress.
They dug a pit in my path—
but they have fallen into it themselves.

Between sessions Mike Luttig and I went over various subjects that were proving problematic. My opponents were armed with long lists of trick questions prepared by law professors and activists; I, on the other hand, had spent most of the preceding decade running a federal agency instead of studying two centuries' worth of

Supreme Court decisions, and *Roe* v. *Wade* wasn't the only area of constitutional law about which I'd yet to think deeply. Trying to review so many cases in the space of three months was like trying to cram for a final exam while being shoved around by an angry mob. It wasn't that I doubted my ability to master the material. I already understood the key cases and the legal concepts behind them perfectly well. But it's one thing to know a precedent and another one to think it through methodically, then apply it to specific cases. Until he's gone through that deliberative process on a case-by-case basis, an open-minded judge can't predict how he will rule in any given situation. As for the matter of my judicial philosophy, I didn't have one—and didn't want one. A philosophy that is imposed from without instead of arising organically from day-to-day engagement with the law isn't worth having. Such a philosophy runs the risk of becoming an ideology, and I'd spent much of my adult life shying away from abstract ideological theories that served only to obscure the reality of life as it's lived.

On the other hand, I now saw that there was no reason for me to worry about my ability to discuss such broad-brush questions with the easy fluency of a Robert Bork. Most of my opponents on the Judiciary Committee cared about only one thing: how would I rule on abortion rights? I knew it was irresponsible for them to expect me to prejudge a complex area of law without having decided a single case on the subject, but I also knew that it had never occurred to any of them that my personal view about the morality of abortion would have nothing to do with my view of *Roe* v. *Wade*. I wasn't that kind of judge—or that kind of person. I had sworn to administer justice "faithfully and impartially." To do otherwise would be to violate my oath. That meant I had no business imposing my personal views on the country. Nor did I have the slightest intention of doing so. From the start of my tenure on the court of appeals, I'd taken Larry Silberman's advice to heart: in every case that came before me, I considered what my role was *as a judge.* But my enemies weren't looking for open-minded justices. All they cared about was keeping anyone off the Supreme Court who might possibly vote to reverse *Roe* or water it down. As far as they were concerned, my open-mindedness was a disadvantage, not a qualification.

I spent five days testifying before the Judiciary Committee. When I was through, I couldn't shake the feeling that for all the intensity of their effort, my opponents were still holding something in reserve. But I couldn't imagine what it might be, and after two and a half months of constant preparation and unrelenting attacks, I desperately needed a break. As soon as I concluded my testimony, Virginia and I drove to the eastern shore of Maryland, caught a ferry to New Jersey, and rented a tiny room in Cape May, an old-fashioned resort town on an island located at the southern tip of New Jersey. The summer tourist season was over and the beach was deserted and quiet, but I found no peace there. All I could think about was what I'd just been through—and what lay ahead. Being away from Washington was just as bad as being there.

As bad as I felt, though, my mother felt even worse. Between the day President Bush announced his intention of nominating me to the end of my testimony, she lost more than thirty pounds as a result of stress and worry. A lifelong Democrat who had always admired the Kennedys, she grew increasingly furious with the Democratic senators who were trying to sabotage my nomination, though the unceasing attacks on me had already taken their toll on her by the time the hearings started. I'd told her to be careful about talking to reporters, but that hadn't stopped her from granting countless interviews, since she was sure that it couldn't hurt to tell our story. Within a few weeks, she knew better. One reporter actually had the nerve to argue with her about how many children she had, insisting that there were only two of us.

Leola and I had never before discussed political matters. Daddy had once asked me why I'd become a Republican, to which I replied that the Democrats no longer represented the things he'd taught me. But I never asked my mother how she voted, nor did she ask me why I'd chosen to ally myself with a party that so many blacks regarded as racist and evil. Now she could see for herself. Patrick Leahy, Howard Metzenbaum, Joe Biden, Paul Simon, even Teddy Kennedy: all of them were arrayed against me. How dare they treat her son that way. Never before had I seen her as angry as she was in the fall of 1991. All her life she'd assumed that Democrats in Washington were sensible leaders—but now she saw these men as single-issue zealots who were unwilling to treat her son fairly. "I ain't never votin' fo' another Democrat long as I can draw breath," she told me as we walked out of the Senate building on what should have been my final day of testimony. "I'd vote for a *dog* first."

9

INVITATION TO A LYNCHING

The day after Virginia and I returned from Cape May, Lee Liberman, a lawyer in Boyden Gray's office, called to tell me that the FBI needed to send two of its agents to Alexandria to speak with me as soon as possible. "I can't tell you what it's about," she said, "except that it's something silly—but we'd better play it safe. When can they come to see you?"

"The sooner, the better," I said. "Waiting will only make it worse."

I called Mark Paoletta as soon as Lee hung up. He and I had been speaking several times each day, and until now we'd been completely open with each other. No more. "I can't tell you anything," he said meekly. "The FBI will tell you what this is all about." His voice sounded unnatural, as though he were deliberately restraining himself and hated having to do so.

After talking to Mark, I paced around the room like a caged animal. Lee had made it sound as though there was nothing to worry about—but Mark's awkward, closemouthed response told a different story. Had some disgruntled employee at EEOC come to the FBI at the last minute to lodge a complaint against me? I had no idea what to expect, but I feared the worst.

The phone rang again. This time it was my friend Butch Faddis, calling from St. Louis. "I'm flying to Washington," he said. "I can't stand watching you get beat up from a distance. I need to spend some time with you. I'll be there this afternoon." I was too distracted to say much of anything, so I thanked him and hung up.

Two FBI agents came to the house later that morning, a black man and a white woman. They flashed their credentials and started asking questions before I could close the door behind them.

"Do you know a woman named Anita Hill?"

"Yes, of course." I was astonished. What interest could the FBI possibly have in Anita?

"Did you ever make sexual advances to her or discuss pornography with her?"

"Absolutely not," I said. I couldn't believe what I was hearing. Carlton Stewart, a childhood friend who'd worked with Anita at EEOC, had told me he'd run into her at the ABA convention in August and that she'd been delighted by the news of my nomination.

I walked the agents into the kitchen. We sat down at the table and they took out a document, but wouldn't show it to me. It was, they said, a copy of a written statement that Anita had given to the Judiciary Committee. (She had initially requested that her name be withheld from the members, but Senator Biden, to his credit, refused to consider it unless it was signed.) The parts they read aloud made vague, unsupported allegations of some unspecified sexual misconduct by me. In support of her charges, Anita had offered the names of several members of my personal staff at EEOC as corroborating witnesses—none of whom, I noted with surprise, had gotten along with her. The agents asked whether she'd ever contacted me after leaving the agency. Apparently Anita was claiming that she had done so only in response to my calls. I said that in fact she had called my office regularly, and that I called her back whenever she had a reason to speak to me directly.

I told them the story of how Gil Hardy had asked me to give Anita a job ten years earlier, and that her work at EEOC had been mediocre. They asked if I'd wanted to date her. "My goodness, no!" I said. The question would have been laughable if the situation hadn't been so deadly serious. The agents said that Anita's allegations were being investigated across the country even as we spoke, and that the FBI report would be finalized by the end of the day. If any follow-up questions needed to be asked, they'd get back to me that afternoon. Then they departed, leaving me shaken and demoralized. The interview had taken less than an hour.

I called Virginia's office immediately. She asked if there was anything more that she needed to know about my relationship with Anita Hill. "Nothing," I said. I explained how I'd hired her at Gil's urging and done my best to help her, just as I'd tried to help so many other young blacks; I described how Anita had stormed out of my office when I promoted Allyson Duncan instead of her, complaining that I preferred light-complexioned women. Virginia said she loved me more than ever and promised to stick with me through thick and thin, but I could hear the pain in her voice. I'd warned her that some of my opponents would try to kill me. Now we both knew what their weapon of choice was to be: the age-old blunt instrument of accusing a black man of sexual misconduct. And it did not matter that a black woman was being used to make the accusations.

Afterward I sat in silence, thinking through what had just happened. I'd been waiting for the other shoe to drop, and now it had. But just what was Anita specifically claiming that I'd said or done—and, just as important, why was she claiming it? I felt sure that I had never said or done anything to her that was even remotely inappropriate, but I knew that in Washington, what matters is not what you do but what people can be made to think you've done. I also knew from working with Anita that she was touchy and apt to overreact. If I or anyone else had done the slightest thing to offend her, she would have complained loudly and instantly, not waited for a decade to make her displeasure known. Of course, we'd disagreed sharply about politics—I remembered how she'd said at our first meeting that she "detested" Ronald Reagan—and I'd found her political views to be both stereotypically left of center and uninformed. But I never allowed political differences with my subordinates at EEOC to stop me from working cordially with them, and Anita was no exception. Outside of purely political matters, the only thing about which we'd argued during the time we worked together was my refusal to promote her, and even that hadn't stopped me from helping her get another job. Why, then, was she now attacking me in a way that seemed calculated to do the maximum possible damage at the worst possible time?

I was one of the least likely candidates imaginable for such a charge. I'd always insisted on a professional and respectful workplace, going to great lengths to discourage fraternization among my personal staff. I wanted to show that a predominantly minority and female agency could be run as professionally as any other—and that it could be done without the benefit of affirmative action or quotas. All that was necessary, I believed, was a concerted effort to give those who had been excluded an opportunity to do their best. I'll always remember running into a black EEOC staffer on the very day she'd been promoted to a senior management position. Her panic was obvious: she told me in breathless bursts of words that she knew she could do her old job but wasn't so sure about the new one. She had only a high school education and had never envisioned being elevated to so high a post. "Don't worry," I said. "You can have your old job back if this one doesn't work out, but I'm sure you'll do just fine. We didn't promote you because you're a black woman. We did it because we thought you deserved the job." So she did, and she lived up to our expectations.

Nor did I hold myself to a lesser standard than my employees. I was proud—perhaps too much so—that throughout a decade of increasingly contentious relations with civil-rights groups and the press, no one had ever accused me of any personal impropriety. Quite the contrary: my reputation was spotless. In 1987 EEOC had held a series of training sessions in Dallas for its investigators, and one of them, a black woman who'd been working at the agency since before I became chairman, made a point of telling me that she appreciated the professional manner in which I'd always conducted myself, adding that she'd never heard a single rumor about my personal life. This was no accident. I'd grown up hyperconscious of the need to avoid even the appearance of such impropriety, for I was intensely aware of America's long and ugly history of using lies about sex as excuses to persecute black men who stepped out of line. Reading Richard Wright's *Native Son* had made the strongest possible impression on me as a college student. What had happened to Bigger Thomas, I knew, could happen to any black man, including me. "To hint that he had committed a sex crime was to pronounce a death sentence. . . ." This was why I'd gone out of my way to avoid the very behavior of which I was now being accused.

That afternoon one of the FBI agents who'd visited me called to say that the report was complete and that all of Anita's corroborating witnesses whom they had mentioned to me had steadfastly denied her allegations. I was relieved, but not for long. Butch Faddis arrived that afternoon. We went back a long way—he had talked me into running the Marine Corps Marathon—and I knew I could trust him. We drove to my favorite spot in Mount Vernon, and I told him what had happened as we sat looking at the Potomac River. He laughed out loud. "That's ridiculous," he said. "You know it won't go anywhere. People will see right through it." I explained that I wasn't going before a judge and jury. Instead I was up against a phalanx of smart, well-heeled interest groups that were working hand in hand with the media and the powerful politicians who opposed my nomination. They were out to kill me, I said, and they'd stop at nothing. Butch scoffed at my pessimism—but I knew better.

Leola called that evening. I didn't have the heart to tell her about Anita's charges, especially since she had good news: her first great-grandchild, Mark Martin, had just been born. I was too obsessed with my own crisis to share her joy. Little did I know that Mark's birth would later seem like a desert flower that had bloomed in the midst of the barren landscape of my confirmation. He was the first-born child and namesake of my sister's youngest son, and six years later, Virginia and I took him in, just as my grandparents had taken Myers and me into their home and changed forever the course of our lives.

For the moment, the Judiciary Committee kept Anita's statement under wraps. Senator Danforth assured me that so far as he could tell, it hadn't changed a single vote on the Judiciary Committee, whose members were hesitant to rely on any allegation that the FBI couldn't substantiate. But another surprise was on the way. I spoke to the FBI agents on Wednesday; someone at the court of appeals leaked one of my draft opinions to the press on Friday. This breach of confidentiality was unprecedented. One of the hallmarks of the federal judiciary had always been the absolute secrecy in which it worked. Leaks were unthinkable—until now. The case in question involved preferences given to women by the Federal Communications Commission in awarding radio-station licenses, and it was clear that my opinion had been leaked by a person or persons who wanted to portray me as unsympathetic to women's causes. The case had been heard by a three-judge panel whose other members were Jim Buckley and Abner Mikva. Jim and I had agreed that the preference was unconstitutional, with Judge Mikva dissenting. While final opinions are circulated to the full court, early drafts are shared only among the members of the panel hearing the case, and it would have been highly unusual for anyone other than those judges and their law clerks to have access to them. I knew nobody in my office was responsible for the leak, and I was confident that it had not come from Judge Buckley or any of his staff.

I was appalled. Until then I'd found the court of appeals to be an oasis of calm and mutual trust, a place that had made it possible for me to escape from the trumped-up controversies and political gamesmanship of my life at EEOC. I'd been free to do my work without having to constantly watch my back. Now those days were over, and I feared that even if I were confirmed, life at the Supreme Court would be no more peaceful. But I'd promised President Bush that I could make it through another confirmation, and I couldn't go back on my word. I'd done that only twice in my life, once with Daddy and once with Kathy, and I wasn't about to do it again. As always, it was the memory of Daddy that strengthened me. "Son, you have to stand up for what you believe in," he had said. "Give out, but don't give up." I knew I couldn't give up—I couldn't live with myself if I did—but I feared that I might give out.

The Judiciary Committee was set to vote on my nomination on Friday. Senator Arlen Specter, a Republican member of the committee, had asked to speak with me privately first, and the two of us met in Senator Danforth's office on Thursday. He started out by telling me that he'd read the FBI's report on Anita Hill's allegations, and that her statement had left him with the impression that Anita thought I was interested in her romantically, since I hadn't been dating anybody at the time. That simply wasn't true, I explained: I'd dated the same woman throughout the time Anita had been working for me. Senator Specter seemed satisfied with my assurances, but by then I felt like a character out of Franz Kafka's *The Trial*. "Someone must have been telling lies about Josef K.," Kafka's novel begins, "for without having done anything wrong he was arrested one fine morning." Not only had I done nothing wrong, but I didn't even know what I was supposed to have done.

Ken Duberstein, a Washington lobbyist who had volunteered to help steer me through the confirmation process, called the next morning to say that Joe Biden wanted to talk to me before the vote. I called the Judiciary Committee cloakroom, and after a brief wait, Senator Biden came on the line. I held the receiver sideways so that Virginia could hear him speak as we stood together in the kitchen. The senator said that he was torn over his decision and had actually brought two statements with him to the committee meeting that day, one for me and the other against. He had decided to oppose me. He'd voted to confirm Justice Scalia, he explained, and now regretted it; he thought it was possible that I might turn out like Justice Scalia, so he couldn't vote for me.

"That's fine," I said. "It doesn't matter to me whether I'm confirmed or not. But I entered this process with a good name, and I want to have it at the end."

"Judge, I know you don't believe me," he replied, "but if any of these last two matters come up, I will be your biggest defender." (The other matter to which he was referring was the leak of my draft opinion.)

He was right about one thing: I didn't believe him. Neither did Virginia. As he reassured me of his goodwill, she grabbed a spoon from the silverware drawer, opened her mouth wide, stuck out her tongue as far as she could, and pretended to gag herself.

As soon as I hung up, Virginia and I drove to Fork Union Military Academy to watch Jamal play football. Later in the day the Judiciary Committee cast its vote. It was a tie: seven senators voted to send my name to the full Senate with a positive recommendation, and seven voted not to recommend me. One Democrat, Dennis DeConcini of Arizona, courageously broke ranks to support me. Otherwise it was a straight party-line vote. "For this senator, there is no question with respect to the nominee's character, competence, credentials, or credibility. . . . This is about what he believes, not about who he is," Senator Biden fulsomely proclaimed before voting against me.

The full Senate vote was scheduled for Tuesday, October 8, a week and a half away. So far as I knew, there was nothing more I could do but wait—and pray. I had been reading Bible verses all summer, and as the confirmation process ground on, I spent even more time doing so. I knew that many good people were working tirelessly to help get me confirmed, but that knowledge no longer calmed my nerves or lifted my spirits. The more hopeless things appeared and the more vulnerable I felt, the more I turned to God's comforting embrace, and over time my focus became primarily God centered. The words of the apostle Paul were never far from my mind: "Therefore I take pleasure in infirmities, in reproaches, in necessities, in persecutions, in distresses for Christ's sake: for when I am *weak*, then am I *strong*."

Many people now supposed that the worst was over. Justice David Souter, for instance, sent a handwritten note congratulating me and welcoming me to the Supreme Court family. Chief Justice Rehnquist's administrative assistant, Rob Jones, called to offer assistance in setting up my chambers. Ken Duberstein called the next Friday, a week after the committee vote, sounding almost giddy. The battle, he said, was over: we had between seventy and eighty votes lined up. Knowing that I'd kept my Corvette in the garage all summer to avoid unnecessary scrutiny, Ken told me I could now take it out for a drive. I declined. I didn't trust my enemies—or, for that matter, the commitment of all those seventy or eighty senators. I knew I wouldn't rest easy until the votes were tallied.

I WAS RIGHT to be dubious. Lee Liberman called the next night to tell me Anita's supposedly confidential statement had been leaked to Nina Totenberg of NPR and Tim Phelps of *Newsday* and would become public in a matter of hours. Virginia and I tried frantically to get in touch with Ken Duberstein, but he was at a baseball game.

On Sunday morning, courtesy of *Newsday*, I met for the first time an Anita Hill who bore little resemblance to the woman who had worked for me at EEOC and the Education Department. Somewhere along the line she had been transformed into a conservative, devoutly religious Reagan-administration employee. In fact she was a left-winger who'd never expressed any religious sentiments whatsoever during the time I'd known her, and the only reason why she'd held a job in the Reagan administration was because I'd given it to her. But truth was no longer relevant: keeping me off the Supreme Court was all that mattered. These pieces of her sordid tale only needed to hold up long enough to help her establish credibility with the public. They fell away as the rest of the story gained traction in the media, just as the fuel tank and booster rockets drop away from a space shuttle once it reaches the upper atmosphere.

I was struck by the glaring difference in the way the media treated Anita and me. Whereas it was taken for granted that whatever she said had to be true, it was no less automatically assumed that anything I might say in my defense would be untrue. The same kind of reflexive groupthink had been at work when my leadership of EEOC was portrayed as "controversial" before I'd had a chance to do anything there. What made it controversial, of course, was that I refused to bow to the superior wisdom of the white liberals who thought they knew what was better for blacks; since I didn't know my place, I had to be put down. The same process, I recognized, was at work now. I had never feared the results of the FBI's investigation, not merely because I was innocent but also because I trusted the agents to behave professionally. What I feared was that if Anita's charges became public, the media would jump to its usual conclusion that I was the villain—and that was what happened.

On Monday I went to the White House to meet with Senator Max Baucus, a Democrat from Montana who clearly had no intention of voting for me, though he went through the motions of asking a string of superfluous questions. Meanwhile Anita and her supporters were staging their first press conference in Oklahoma. Virginia watched it on TV, and later told me it had been a slickly choreographed event in which Anita was presented as a pious young woman who had mustered up the courage to come forward and confront the powerful villain who had mistreated her. The possibility that she might have been motivated not by personal outrage but by ideological conviction was broached by none of her obliging questioners. Like Senator Baucus's visit with me, it was all for show.

Up to that moment, I'd somehow managed to keep hold of my emotions, but no sooner did Senator Baucus leave than I lost my grip. I'd met with him in Fred McClure's office, and now I ran into the hallway, wild-eyed and desperate, demanding to talk to Fred at once. I started spewing rambling questions at him. How could someone I'd tried to help turn on me so viciously? What was I supposed to do now? How could I prove a negative—especially when a vast army of political operatives, left-wing academics, public-relations firms, and cynical reporters was covering me with slime? Fred had no answers, and neither did Virginia when I called her at work, half crazed with fear. Ever since Daddy and Aunt Tina died, I'd done my best to live a life that would be worthy of their sacrifices. It was to be my memorial to them, and now it had been besmirched. After a lifetime of struggle and achievement, I'd been thrust back into Bigger Thomas's world, a dark, cramped hell devoid of hope:

> He felt he had no physical existence at all right then; he was something he hated, the badge of shame which he knew was attached to a black skin. It was a shadowy region, a No Man's

> Land, the ground that separated the white world from the black that he stood upon. He felt naked, transparent; he felt that this white man, having helped to put him down, having helped to deform him, held him up now to look at him and be amused.

I was spent by the time the U.S. marshals drove me home from the White House. I felt like a marathon runner who had hit the wall. All my reserves were used up. I lay across the bed and curled up in a fetal position, tired beyond imagining. More than anything else in the world, I wanted to go back to Georgia, where life had been simple and honest. It wasn't just the events of the past few weeks that weighed on me: I felt myself crushed beneath the accumulated trials of a lifetime. Never before had I felt their full effect, and the burden seemed more than I could bear. Ever since leaving home, I'd played by the rules—but where had it gotten me? Whites could change those rules whenever they pleased. It had always been that way, and always would be. I'd graduated from one of America's top law schools—but racial preference had robbed my achievement of its true value. I'd been nominated to sit on the Supreme Court—but my refusal to swallow the liberal pieties that had done so much damage to blacks in America meant that I had to be destroyed. Perhaps I had known that all along.

BUT EVEN AS I entered my dark night of the soul, my friends and colleagues kept on fighting to clear my name and keep my spirits up. The phone continued to ring constantly, only now the voices on the other end of the line all wanted to help. People who had worked with me at EEOC called by the score to say that they knew I'd done nothing to Anita Hill, and that the meek and humble young woman they'd seen on TV was nothing like the abrasive, ambitious person they had known. Some didn't even bother to call, but simply caught the next plane to Washington. Not all of them agreed with my political views, but every one of them was sure that Anita's story was a pure fabrication, and they were determined to right the wrong she'd done. They participated in news conferences, submitted written statements to the Judiciary Committee, and gave interviews to the few reporters who were willing to hear them out. Mike Berman had been right: I'd done all I could, and now it was time to let my friends defend me.*

Virginia and I drove to Ricky and Larry Silberman's Georgetown home late that night to escape the media circus that we knew would descend on our house when the full Senate voted on my confirmation the next day. I came downstairs on Tuesday morning to find Larry reading the newspapers and watching *Today* and *Good Morning, America.* Virginia and Ricky left to do what they could to help, while Larry stayed home from the courthouse to keep me company. I spent most of the day pacing around the Silbermans' pool, smoking cigars and worrying, though the two of us took time out to draft a sworn statement in which I unequivocally denied all of Anita's charges. Late in the afternoon I talked on the phone with Jack Danforth, Orrin Hatch, and Bob Dole, who told me that unless the Judiciary Committee was reconvened to examine the new charges, my nomination would be defeated. None of them bothered to bring up the obvious fact that someone connected with the committee had broken the law by leaking Anita's confidential statement to the press: we all knew that to dwell on legal niceties was pointless. I re-

* Virginia and I had already been receiving tokens of support—encouraging letters and telegrams, flowers, cards, Bibles, and other gifts—from total strangers, but what started out as a trickle now became a flood. In addition, our neighbors did everything they could to help, mowing our grass, watering our flowers, and stopping by regularly with food. Having seen so much of the worst of people all summer long, it was rejuvenating to find ourselves on the receiving end of these unsolicited acts of kindness.

luctantly agreed to postpone the Senate vote and face my tormentors once more—to clear my name, if for no other reason.

Virginia and I went home to Alexandria to find our answering machine full of reassuring messages. We spent the evening praying, reading the Bible together, and listening to religious music. Before going to bed, we asked four of our friends, Elizabeth and Steven Law and Kay and Charles James, to come over the next morning and join us in prayer. They showed up bright and early, carrying bags of doughnuts and bagels past the reporters camped outside the house. The six of us chatted for a little while, then sat in a circle, held hands, and asked the Lord for help. Both couples came back each day until the battle was over, and their company was a priceless gift. "Where two or three are gathered in my name," Jesus said, "I am there among them." He was among us now.

It had long since become clear to me that this battle was at bottom spiritual, not political, and so my attention shifted from politics to the inward reality of my spiritual life. I had been proud of my work at EEOC and the Department of Education, and no less proud that I'd spent nearly a decade in the public eye without being touched by personal scandal. Might I have been *too* proud? It occurred to me for the first time that I had cherished my good name in the same way that a wealthy man cherishes his money. I remembered how Jesus had told the rich man to give away his fortune and "come and follow me." Perhaps I would have to renounce my pride to endure this trial, even as Cardinal Merry del Val had prayed for deliverance in his Litany of Humility: *Deliver me, O Jesus, from the fear of being humiliated . . . from the fear of being despised . . . from the fear of suffering rebukes . . . from the fear of being calumniated.*

In addition to suspecting that I had committed the sin of pride, I saw that I was resisting what God had put before me. "Father, let this cup pass away from me," Jesus had prayed in the garden of Gethsemane. "But thy will, not mine be done." The second half of

His prayer is the harder part. Until then I'd been concentrating on wanting the confirmation debate to come to an end, drawing back from total submission to God's will. Now I had no choice but to submit completely: I could do nothing to push the cup away. The time had come to attend to His will, not mine. I could not know whether doing so would make the experience less difficult, but I had faith that His transcendent purpose would sustain me to the end of it—and beyond. He had never failed me. Even in my darkest hours, even when I openly rejected Him, His forgiving and sustaining Grace had always been there. I knew that it would give me the glimmer of hope I needed now more than ever. It was in the consoling words of the prophet Isaiah that I found my own watchword: "But they that wait upon the Lord shall renew their strength; they shall mount up with wings as eagles; they shall run, and not be weary; and they shall walk, and not faint."

Senator Danforth drove out to the house later that morning to talk with Virginia and me. "I don't know how this is going to turn out, Clarence," he said. "But don't forget, you never wanted to be on the Court." I agreed. I was ready to do my best—I owed it to the president, to my supporters, and to everyone else who had been similarly mistreated in the past, starting with Judge Bork—but the Court itself I could take or leave. It was in God's hands now. The senator drove us into Washington and dropped us off at the White House. A press car followed us all the way there. Virginia and I went to the Oval Office and met with President Bush, after which she and Mrs. Bush went off by themselves while the president and I went for a walk on the South Lawn. He said he was sorry for having gotten me into so dirty a fight. I said that I didn't blame him in the slightest. He made it clear that he wouldn't abandon me, and I knew he meant it. "I'll do my best to stick it out," I said. "I promised I would." Later I learned that President Bush's team had begun to fracture. Marlin Fitzwater was among the White House staffers who talked of pulling the plug on my nomination, but Boyden Gray—as well as the president himself—refused to panic.

From there I went to the courthouse, where I took calls from several of my colleagues, all of whom expressed their outrage at the way I was being treated. Judge Aubrey Robinson was particularly angry with my persecutors. "I would just tell them all to go to hell," he said. Larry Silberman gave me a more practical piece of advice: get a lawyer. Until that moment it hadn't crossed my mind that I might need one. None of my advisers, I realized, was a trial lawyer. Larry gave me some names, but as much as I trusted his judgment, I also knew that what I really needed was a good friend who knew his way around a courtroom, so I called Larry Thompson that evening. Larry had done well for himself since we'd worked together at Monsanto, serving as the U.S. Attorney for the Northern District of Georgia during the Reagan administration, then becoming a partner at King and Spalding, a prestigious Atlanta law firm with a thriving white-collar criminal practice. Since the Judiciary Committee was treating me like a common criminal, he was just the lawyer I needed.

"Larry, I need your help," I said.

"I'll be there on Monday."

"It'll all be over by then."

"Then I'll be there in the morning." And that was that.

The hearings would be reopened on Friday. That gave Larry a day to think things through. He didn't need to do much thinking, though, for Anita's public statements, as Lee Liberman had hinted when she'd first called me from the White House, were full of holes. She'd claimed at her press conference to have been too afraid of me to complain about my alleged misconduct—yet she'd lobbied aggressively to follow me from the Department of Education to EEOC. She said she'd never called me—but the telephone logs of my secretaries at EEOC and the court of appeals proved that she'd done so repeatedly. She claimed that other members of my staff could corroborate her story—but they denied it. In the end only three EEOC employees would support her version of what supposedly happened between us—but all of them had either been fired or left the agency on bad terms, and none, to my knowledge, had worked there at the same time as Anita. Having spent years at EEOC reviewing such claims, I was sure that this one would have been thrown out of court in an instant. But did any of these things matter? Not in the least. The mob was howling, and it wouldn't be satisfied until it had tasted my blood.

The more I reflected on what was happening, the more it astonished me. As a child in the Deep South, I'd grown up fearing the lynch mobs of the Ku Klux Klan; as an adult, I was starting to wonder if I'd been afraid of the wrong white people all along. My worst fears had come to pass not in Georgia but in Washington, D.C., where I was being pursued not by bigots in white robes but by left-wing zealots draped in flowing sanctimony. For all the fear I'd known as a boy in Savannah, this was the first time I'd found myself at the mercy of people who would do whatever they could to hurt me—and institutions that had once prided themselves on bringing segregation and its abuses to an end were aiding and abetting in the assault. Hypersensitive civil-rights leaders who saw racism around every corner fell silent when my liberal enemies sneered that I was unqualified to sit on the Court; editors and reporters who claimed to be objective substituted a pretense of balance for true fairness, presenting outrageous, wholly unsupported allegations side by side with sputtering denials. The implausible was now being treated more favorably than the obvious.

As for the Senate, it had abandoned all semblance of decorum to consider a set of trumped-up charges better suited to the tabloids than the *Congressional Record*. I knew of at least one senator sitting in judgment of me against whom accusations of sexual improprieties had been leveled that made Anita's charges look mild. I wondered how many others had been trapped by their own failings into ignoring the obvious weaknesses of her claims. I was as sickened by their hypocrisy as I was mystified by the sequence of events that had set this hideous farce in motion.

Boyden Gray called on Thursday night with one final question. Did I want to testify first on Friday, or let Anita precede me? I said I'd do whatever he recommended. He suggested that I testify first, and reminded me to prepare a statement. I had not thought to do so. I started hyperventilating, and I could feel my heart pounding in my chest. Virginia tried to distract me by suggesting that we ask Marie, a hair stylist who lived next door, to give me a haircut. I hadn't had one all summer, so I agreed. Marie and her husband, David, came over around nine o'clock. Their presence briefly took my mind off my troubles and my sheer exhaustion, but as soon as they left, I was overcome by the feeling of constant dread that had been with me all summer and now was more intense than ever before. I thought of all the mistakes I'd made in my life. I shouldn't have left Savannah. I shouldn't have left the seminary. I shouldn't have gotten divorced. I shouldn't have given so many speeches, or expressed myself so bluntly. But try as I might, I couldn't see what I'd done wrong with Anita, except to have hired her in the first place.

I had been sleeping badly all summer and in recent days hardly at all, so I went upstairs to our bedroom and lay down to rest for a little while, drifting in and out of consciousness. I felt sand, hot and dry, on the bottom of my feet. The dark leaves of magnolia trees waved back and forth in the breeze. Palm trees stood in bold relief against the backdrop of the marshy savannah. The hot, moist salt air soothed my lungs like a steam bath. I ran and played, skipping flat oyster shells off the murky salt water in the creek. I was a boy again, back home in Pinpoint, safe and secure . . .

"Honey? Honey?" I heard a soft voice calling me from out of the breeze. "Honey, it's after midnight. You have to get up and write your statement." It was Virginia. I'd dozed off for a few minutes, but now I had to get up and face the horrible nightmare that my life had become. I fought my way back to consciousness, then went back downstairs to the kitchen and sat at the table, which was covered with papers. I stared at the clutter, unable to summon the energy to do anything. "Clarence, you have to get started," Virginia said gently. "Here are some suggestions from Nancy Altman." Nancy was an old friend with whom I'd worked in Senator Danforth's office and at the Department of Education. Like so many of the other people I knew, she'd offered her help as soon as the story broke. Now Virginia handed me a piece of paper on which she had scribbled down some of Nancy's ideas.

"I'm confused, Virginia," I said helplessly. "All these papers on the table have me confused." She swept them away and placed a legal pad in front of me. *Lord, give me the wisdom to know what to write and the courage to write it,* I prayed. Then I picked up my pen, and all at once the words began to pour out of me. I wrote for four straight hours. Virginia took each handwritten page upstairs and typed it into the computer, editing as she went. Then she printed out a complete draft and we edited it together, finishing around five in the morning. By then we were both punch-drunk with exhaustion—emotional, mental, and physical. We decided to lie down until six, then call Senator Danforth.

I spent the hour tossing, turning, and thinking, and the more I thought, the angrier I got. As a child I'd labored in the South Georgia heat because, Daddy said, it was our lot to work from sun to sun. I'd lived by the rules of a society that had treated blacks shabbily and held them back at every turn. I'd plugged away, deferred gratification, eschewed leisure. Now, in one climactic swipe of calumny, America's elites were arrogantly wreaking havoc on everything my grandparents had worked for and all I'd accomplished in forty-three years of struggle. Should I have seen it coming? Even as Daddy had been teaching me that hard work would always see me through, my friends in Savannah told me to let go of my foolish dreams. "The man ain't goin' let you do nothin'," they had said over and over. "Why you even tryin'?" Now I knew who "the man" was. He'd come at last to kill me, and I had looked upon his hateful, leering face as he slipped his noose of lies around my neck.

Twenty years earlier I'd prayed to God to purge my heart of anger, and since then I had managed to hold the beast of rage at bay. Now it had slipped its leash—but for a very different reason. I didn't care whether I ever sat on the Supreme Court, but I wasn't going to let what little my family and I had cobbled together be so wantonly smashed. My enemies wanted nothing more than for me to go quietly. I, on the other hand, owed it to my family and the memory of my grandparents and forebears not to self-destruct but to confront them with truth.

At six o'clock Virginia and I got out of bed. The flame of anger burning inside me had done its work. I called Senator Danforth and read him my statement. He made only two suggestions, both of which I accepted: I cut out the usual expressions of gratitude with which witnesses at Senate hearings open their statements, and deleted a reference to the conversation in which Joe Biden had assured me that he'd be my "biggest defender" if Anita Hill's charges became public. That, Senator Danforth said, would be counterproductive. Virginia printed out the final draft of the statement, and we dressed, prayed, and left for Capitol Hill. The man was waiting for me there—and this time, with God's help, I would be ready for him.

10

GOING TO MEET THE MAN

Virginia and I went straight to Senator Danforth's private office. I pleaded with him to keep everyone else away from me. I was grateful for their help, but they were interested in getting me confirmed, and that no longer mattered to me. The senator led us in prayer, exhorting me once again to let the Holy Ghost speak through me. Then we went to the hearing room. I walked directly to my seat, not bothering to greet the members of the committee with the ritual cordiality expected of a witness. I was astounded by how diminished they seemed. A month ago I had thought them larger than life, but now they were small and irritating. I didn't care what they did anymore: they had nothing that I wanted. I had come to speak for myself and my family, to tell the world the truth about the monstrous thing that had been done to us.

As I started to read my statement, the room seemed oddly quiet, and the turmoil inside me died away. I was at peace.

"As excruciatingly difficult as the last two weeks have been," I said, "I welcome the opportunity to clear my name today. No one other than my wife and Senator Danforth, to whom I read this statement at six-thirty a.m., has seen or heard this statement. No handlers, no advisers.

"The first I learned of the allegations by Professor Anita Hill was on September 25, 1991, when the FBI came to my home to investigate her allegations. When informed by the FBI agent of the nature of the allegations and the person making them, I was shocked, surprised, hurt, and enormously saddened. I have not been the same since that day. . . .

"I have been racking my brains and eating my insides out trying to think of what I could have said or done to Anita Hill to lead her to allege that I was interested in her in more than a professional way and that I talked with her about pornographic or X-rated films. Contrary to some press reports, I categorically denied all of the allegations and denied that I ever attempted to date Anita Hill when first interviewed by the FBI. I strongly reaffirm that denial."

I spoke of my professional relationship with Anita. I said nothing of how her work at EEOC had been mediocre, and made no mention whatsoever of the awkward circumstances under which she'd left Washington to join the faculty of Oral Roberts University. All I said was that I'd recommended her highly to Dean Kothe and told him that I thought she'd make a good teacher, adding that she'd called my office on occasion after leaving the agency.

I continued:

"Throughout the time that Anita Hill worked with me, I treated her as I treated my other special assistants. I tried to treat them all cordially, professionally, and respectfully and I tried to support them in their endeavors and be interested in and supportive of their success. I had no reason or basis to believe my relationship with Anita Hill was anything but this way until the FBI visited me a little more than two weeks ago. I find it particularly troubling that she never raised any hint that she was uncomfortable with me. She did not raise or mention it when considering moving with me to EEOC from the Department of Education, and she'd never raised it with me when she left EEOC and was moving on in her life. And, to my fullest knowledge, she did not speak to any other women working with or around me who would feel comfortable enough to raise it with me. . . .

"As a manager, I made every effort to take swift and decisive action when sex harassment raised or reared its ugly head. The fact that I feel so very strongly about sex harassment and spoke loudly at EEOC has made these allegations doubly hard on me. I cannot imagine anything that I said or did to Anita Hill that could have been mistaken for sexual harassment.

"But with that said, if there is anything that I have said that has been misconstrued by Anita Hill or anyone else to be sexual harassment, then I can say that I am so very sorry and I wish I had known. If I did know, I would have stopped immediately and I would not, as I've done over the past two weeks, have to tear away at myself, trying to think of what I could possibly have done. But I have not said or done the things that Anita Hill has alleged. God has gotten me through the days since September 25, and He is my judge."

Now I turned to the consequences of Anita's reckless charges:

"Mr. Chairman, something has happened to me in the dark days that have followed since the FBI agents informed me about these allegations. And the days have grown darker as these very serious, very explosive, and very sensitive allegations were selectively leaked in a distorted way to the media over the past weekend. As if the confidential allegations themselves were not enough, this apparently calculated public disclosure has caused me, my family, and my friends enormous pain and great harm. I have never in all my life felt such hurt, such pain, such agony. My family and I have been done a grave and irreparable injustice. . . .

"When I stood next to the President in Kennebunkport being nominated to the Supreme Court of the United States, that was a high honor; but as I sit here before you 103 days later, that honor has been crushed. From the very beginning, charges were leveled against me from the shadows, charges of drug abuse, anti-Semitism, wife beating, drug use by family members, that I was a quota appointment, confirmation conversion, and much, much more. And now, this.

"I have complied with the rules. I responded to a document request that produced over 30,000 pages of documents, and I have testified for five full days under oath. I have endured this ordeal for 103 days. Reporters sneaking into my garage to examine books I read. Reporters and interest groups swarming over divorce papers, looking for dirt. Unnamed people starting preposterous and damaging rumors. Calls all over the country specifically requesting dirt.

"This is not American; this is Kafkaesque. It has got to stop. It must stop for the benefit of future nominees and our country. Enough is enough.

"I'm not going to allow myself to be further humiliated in order to be confirmed. I am here specifically to respond to allegations of sex harassment in the workplace. I am not here to be further humiliated by this committee or anyone else, or to put my private life on display for prurient interests or other reasons. I will not allow this committee or anyone else to probe into my private life. This is not what America is all about. To ask me to do that would be to ask me to go beyond fundamental fairness.

"Yesterday I called my mother. She was confined to her bed, unable to work and unable to stop crying. Enough is enough. . . .

"Mr. Chairman, I am a victim of this process. My name has been harmed. My integrity has been harmed. My character has been harmed. My family has been harmed. My friends have been harmed. There is nothing this committee, this body, or this country can do to give me my good name back. Nothing. I will not provide the rope for my own lynching or for further humiliation. I am not going to engage in discussions nor will I submit to roving questions of what goes on in the most intimate parts of my private life or the sanctity of my bedroom. These are the most intimate parts of my privacy, and they will remain just that, private."

I STOPPED TALKING. The world seemed to go blank. The senators, their aides, the reporters and cameramen, the audience seated behind me: they were there but not there.

After a moment I collected myself and saw that my statement appeared to have taken the Judiciary Committee and its staffers by surprise. Senator Kennedy in particular looked stunned. Had he and his aides expected me to slink away in shame? Obviously they knew nothing about me. In all my years in government, I had never cut and run. Life had never afforded me that luxury. There was an uncomfortable pause, followed by a brief, agitated exchange of arguments among the members of the committee. Then I was asked to leave. The Democratic senators wanted Anita to testify immediately. I departed, with no intention of coming back. I'd said what I'd had to say.

The marshals drove Virginia and me home to Alexandria. She went upstairs to watch Anita's testimony on the TV in our bedroom. I didn't bother. I had no interest in it. I knew she had nothing to say that I needed to hear. I still wondered why she'd chosen to do what she did, though my guess was that a combination of ego, ambition, and immaturity had caused her to let herself be drawn into the effort to destroy me—but that was Anita's problem, not mine. I had tried to help her. She had betrayed me. Now she was on her own.

I paced the floor of our small family room, then went outside and paced around the deck, exchanging an occasional word with a marshal who had stayed behind in the house with us. Then Virginia came downstairs and told me what Anita had said I had done to her. I have no intention of repeating the dirty details here. Suffice it to say that it was a relief to hear them at long last, since there was no truth to them: they were nothing more than an extravagant fiction concocted so as to have the maximum possible impact on the public. For days I had tormented myself trying to remember whether I had made some meaningless offhand comment that Anita had somehow misunderstood, or offended her in some insignificant way that had long ago slipped my mind. Now I knew that I had done nothing—though I also knew that the media would automatically treat everything she had said, however preposterous it might be, as credible. They had long ago made up their minds about me.

Ridiculous though Anita's charges were, it hurt me to know that they were being blared around the world, and Virginia shared that pain with me. She knew better than anyone that I was a very private person who found it difficult to talk about such intimate matters, much less to know that they were being discussed by total strangers. I have sometimes wondered since then whether my enemies were consciously trying to drive the two of us apart. In fact, though, the opposite happened. We had always trusted each other implicitly, and the fiery trial through which we passed had the effect of melding us into one being—an amalgam, as we like to say. Nor did anyone who knew me more than casually have any doubt that I was innocent of Anita's accusations. Far from losing friends as a result of the hearings, I made new ones. I would never forget the many people who stood up and fought for me. I had only one last doubt in my mind: I didn't know how much more of a beating Virginia and I could take.

Senator Danforth called around five o'clock that afternoon and asked me to come back to his office. I didn't see the point—I was through with the Judiciary Committee—but the senator said that he and Orrin Hatch, the ranking Republican on the committee, had to talk to me face-to-face. Once I got there, both men insisted that it was essential for me to go before the committee that very night and respond to Anita's testimony in person. If her story remained uncontested overnight, Senator Hatch explained, it would monopolize the next day's news cycle, as her handlers had obviously planned. The committee was willing to reconvene at eight o'clock to hear me. Would I testify again? I didn't see what good it would do, but I trusted these two men, so I said yes. Suddenly a great weariness came over me. I knew I had to rest, even though there was next to no time left to do so, and I decided not to bother preparing a second statement. What else was there to say?

I told Senator Danforth that I hadn't watched Anita's testimony. He urged me to let Mike Luttig walk me through the details of what she had said. Several other people then chimed in, wanting to assess the damage Anita had done and discuss how I might respond to her specific accusations. I didn't want to hear from any of them. I already knew from Virginia that she had done nothing but tell lies about me; I saw no point in going into them any further, or in talking about strategy at a moment when getting myself confirmed was the last thing on my mind. All that could possibly do was distract and confuse me. What I needed now was to lie down. "Could you please clear the office?" I asked the senator. He asked everyone to leave, shut the door, and dimmed the lights. Only he and Virginia remained. I lay back on the couch, surrounded by the darkness of early evening, drifting in the liminal space between sleep and waking. Time seemed to slow down, then stop.

I stirred ever so slightly from my half stupor and saw Jack Danforth sitting close by, making notes on a legal pad and watching over me like a guardian angel. Once he had told me that there was room at the top for a black man and given me my first job. Later he'd brought me to Washington and stood by me through a decade of contention and controversy. Most other politicians, I knew, would long ago have looked to their own self-interest and dumped me, but he was still by my side. From the beginning I had tried not to let him down, just as I had tried never to shame my family—and I couldn't do so now.

Jack had been writing down possible points for me to make as I dozed fitfully on the couch, and now he read them out loud to me one by one. As he did so, the thoughts that had been running through my head for the past half-hour crystallized into a single phrase. "Jack," I said, "this is a high-tech lynching."

"If that's what you think," he replied, "then say it."

I took his pad and scrawled "HIGH-TECH LYNCHING" under his list of talking points. Then I slumped back on the couch and returned to my thoughts.

Somewhere in the back of my mind, I must have been thinking of *To Kill a Mockingbird*, in which Atticus Finch, a small-town southern lawyer, defends Tom Robinson, a black man on trial for the rape of a white woman. He was lucky to have had a trial at all—Atticus had already helped him escape a lynch mob's rope. The evidence presented at the trial shows that Tom's accuser had lured him into her house, then kissed him, after which he fled. The case against him is laughably flimsy, but in the Deep South you didn't need a strong case to send a black man to the gallows, and it is already clear that Tom will be convicted when Atticus goes before the jury to make his closing argument:

> The witnesses for the state . . . have presented themselves to you gentlemen, to this court, in the cynical confidence that their testimony would not be doubted, confident that you gentlemen would go along with them on the assumption—the evil assumption—that *all* Negroes lie, that *all* Negroes are basically immoral beings, that *all* Negro men are not to be trusted around our women, an assumption one associates with minds of their caliber.

I knew exactly what Atticus Finch was talking about. I, too, took it for granted that nothing I could say, however eloquent or sincere, was capable of overcoming the evil assumptions in which my accusers had put their trust. I had lived my whole life knowing that Tom's fate might be mine. As a child I had been warned by Daddy that I could be picked up off the streets of Savannah and hauled off to jail or the chain gang for no reason other than that I was black. That was why he had always been so reluctant to travel, even inside Georgia, where it was rare for him to venture outside the three contiguous counties that he knew well. Daddy often told Myers and me stories about the terrible fates that befell foolhardy blacks who made the mistake of leaving home, even to visit such nearby towns as Jesup or Ludowici. Such stories were legion in the Deep South, and though many were surely apocryphal, I knew for a fact that some of them were true. (My friend Bob DeShay, who had a summer job selling encyclopedias, was once told by a Statesboro policeman to get out of town by sundown.) In any case, their point was the fear they instilled in southern blacks, a fear that had helped to keep segregation alive. My generation had sought to replace that fear with a rage that proved over time to be intoxicating, empowering, justifiable—and ultimately self-destructive. Yet we never forgot what it felt like to live in fear of the power of a mob.

The mob I now faced carried no ropes or guns. Its weapons were smooth-tongued lies spoken into microphones and printed on the front pages of America's newspapers. It no longer sought to break the bodies of its victims. Instead it devastated their reputations and drained away their hope. But it was a mob all the same, and its purpose—to keep the black man in his place—was unchanged. Strip away the fancy talk and you were left with the same old story. *You can't trust black men around women. This one may be a big-city judge with a law degree from Yale, but when you get right down to it, he's just like the rest of them. They all do that sort of thing whenever they get the chance, and no woman would ever lie about it. What does it matter that Anita Hill's story doesn't add up?* Something *must have happened. Case closed.*

Somewhere within myself I had to find the words to expose this fraud for what it was. I tore the top sheet from Jack's pad and put it in my pocket. Jack exhorted me once again to let the Holy Ghost speak through me. Then we emerged from the near-dark office and returned to the Caucus Room, drinking in the cheers of the supporters who continued to line the halls. "They're angels," Virginia had told me a month ago when I saw them for the first time and asked who they were. Perhaps she'd been right.

I silently repeated Jack's prayer as I waited for the senators to take their seats: *Let the Holy Ghost speak through me.* Then I began.

"Senator," I said, "I would like to start by saying unequivocally, uncategorically that I deny each and every single allegation against me today that suggested in any way that I had conversations of a sexual nature or about pornographic material with Anita Hill, that I ever attempted to date her, that I ever had any personal sexual interest in her, or that I in any way ever harassed her.

"Second, and I think a more important point, I think that this today is a travesty. I think that it is disgusting. I think that this hearing should never occur in America. This is a case in which this sleaze, this dirt, was searched for by staffers of members of this committee, was then leaked to the media, and this committee and this body validated it and displayed it in prime time across our entire nation.

"How would any member on this committee, or any person in this room, or any person in this country like sleaze said about him or her in this fashion, or this dirt dredged up, and this gossip and these lies displayed in this manner? How would any person like it?

"The Supreme Court is not worth it. No job is worth it. I am not here for that. I am here for my name, my family, my life, and my integrity. I think something is dreadfully wrong with this country, when any person, any person in this free country would be subjected to this. This is not a closed room. There was an FBI investigation. This is not an opportunity to talk about difficult matters privately or in a closed environment. This is a circus. It is a national disgrace. And from my standpoint, as a black American, as far as I am concerned, it is a high-tech lynching for uppity blacks who in any way deign to think for themselves, to do for themselves, to have different ideas, and it is a message that, unless you kowtow to an old order, this is what will happen to you, you will be lynched, destroyed, caricatured by a committee of the U.S. Senate rather than hung from a tree."

WHEN I WAS done, my words seemed to hang in the air of the Caucus Room like the smoke from a bomb that had just exploded. Howell Heflin, whose uncle had once graced the Senate as an apologist for the Ku Klux Klan, had evidently been chosen as the point man for his fellow Democrats, and he began by asking me whether I had watched Anita Hill's testimony.

"No, I didn't," I snapped. "I've heard enough lies."

Senator Heflin's friends in the media liked to describe his manner as "courtly," but now it made me think of a slave owner sitting on the porch of a plantation house. My reply seemed to stun him, though his reaction surprised me as much as my words had surprised him. Why would I have watched it? Her story was obviously a last-minute fabrication that should have been dismissed out of hand, and if he hadn't known it at the time, he certainly should have figured it out by now.

Next came Orrin Hatch, who had known me for a decade and needed no proof of my integrity. He asked me in an apologetic tone to respond to the specific allegations that had been made in public for the first time that afternoon. Had I threatened to ruin Anita's career if she talked about sexual comments that I allegedly made to her? No. Was there any reason why she should "ever have been afraid" of me? No. I made it my business to help my former special assistants. I had given her a recommendation for her position at Oral Roberts University, and had no reason to think she harbored any ill feeling toward me. Did I describe my private parts to her? Did I pressure her to have sex with me? Did I talk about pornographic films with her? No, no, no. With every question he asked, I felt dirtier and dirtier, and the only thing that made it bearable was that Senator Hatch was as embarrassed as I was.

I answered questions for an hour and a half. At the end I felt wearier than ever before, and no more hopeful. But shortly after I got back to Senator Danforth's office, Ken Duberstein rushed in to say that he'd just gotten off the phone with the White House. My testimony, he said, had turned public opinion around in a single stroke. The Capitol's phone lines were jammed with callers, the vast majority of whom were overwhelmingly supportive of me. Later on it occurred to me that for the first time since the crisis had begun, I had spoken directly to the public in prime time instead of having my words reflected in the fun house mirror of the evening newscasts. I thanked God for C-Span and its gavel-to-gavel coverage. Ordinary Americans had seen me for themselves, and seen what had been done to me in their name. Most of them didn't like what they saw. The public, it turned out, wasn't as gullible as my attackers had assumed; their "cynical confidence" turned out to be misplaced. They had underestimated the character and judgment of the American people. For a time I had doubted it as well, and now I was relieved beyond words to see that my pessimism was misplaced.

More bizarre revelations were to come when the hearings resumed on Saturday. Orrin Hatch spoke to me briefly before the session began. "I need to know something that may strike you as peculiar," he said. Then he asked if I'd ever seen *The Exorcist* or read the novel on which it was based. I said that while I might possibly have seen some scenes on TV, I hadn't seen the whole movie, and that my taste in fiction ran more to Louis L'Amour. "I want to warn you," he said, "that some of the questions I'm going to have to ask you are going to be lurid." I didn't see how they could be any worse than what had come before.

Once the hearing began, I started out by expanding on some of the things I'd said the night before. "These are charges that play into racist, bigoted stereotypes, and these are the kinds of charges that are impossible to wash off," I said. "And these are the kinds of stereotypes that I have, in my tenure in government, and the conduct of my affairs, attempted to move away from and to convince people that we should conduct ourselves in a way that defies these stereotypes . . . if you want to track through this country, in the nineteenth and twentieth century, the lynching of black men, you will see that there is invariably or in many instances a relationship with sex—an accusation that that person cannot shake off. That is the point that I am trying to make. And that is the point that I was making last night, that this is a high-tech lynching. I cannot shake off these accusations because they play to the worst stereotypes we have about black men in this country."

What I specifically had in mind was Anita's claim that I had used the term "Long Dong Silver" in conversation with her to refer to private parts. That had been the first thing Virginia mentioned to me when she told me about Anita's testimony. Now Senator Hatch asked me about it, noting that "people hearing yesterday's testimony are probably wondering how could this quiet, you know, retiring woman know about something like Long Dong Silver. Did you tell her that?" No, I said, I didn't, nor could I imagine what had put such an idea in her head. But Senator Hatch thought he knew, and he proceeded to tell the world. Speaking with visible embarrassment, he read aloud an excerpt from the court's opinion in *Carter* v. *Sedgwick County, Kansas*, a 1988 sexual-harassment case: "Plaintiff further testified that on one occasion Defendant Brand presented her with a picture of Long Dong Silver—a photo of a black male with an elongated penis." I recalled that after leaving EEOC, Anita had asked me for permission to use our case files for a research project on which she was working. This particular case, it turned out, was from the federal judicial circuit that includes Oklahoma, where she had been teaching at the time. It would have been readily available to anyone doing research into sexual-harassment case law. After the hearings were over, a former EEOC employee wrote to say that he had investigated that very case, and wondered how my enemies had gotten access to his investigative materials.

The only bright moment came when Senator Hatch mentioned an op-ed column by Juan Williams that had appeared in that morning's *Washington Post*. As the senator read it into the record, my heart spilled over with gratitude. I knew that Juan had put his career on the line in order to say what he thought:

> Here is indiscriminate, mean-spirited mudslinging supported by the so-called champions of fairness: liberal politicians, unions, civil rights groups and women's organizations. They have been mindlessly led into mob action against one man by the Leadership Conference on Civil Rights. . . . To listen to or read some news reports on Thomas over the past month is to discover a monster of a man, totally unlike the human being full of sincerity, confusion, and struggles whom I saw as a reporter who watched him for some 10 years. He has been conveniently transformed into a monster about whom it is fair to say anything, to whom it is fair to do anything. President Bush may be packing the court with conservatives, but that is another argument, larger than Clarence Thomas. In pursuit of abuses by a conservative president the liberals have become the abusive monsters.

When the long day was finally over, it was clear that my testimony had been well received and that the tide of opinion was turning definitively. (The opinion polls that came out on Monday would all be tilted decisively—even lopsidedly—in my favor.) Senator Hatch suggested that he, his wife, Elaine, and the Danforths meet Virginia and me for dinner later that evening. I had been staying home for so long that the idea of eating out in public seemed extraordinary—a small but significant step toward regaining my freedom. We accepted with a mixture of apprehension and delight. A throng of friends and supporters cheered us as we walked out of the Russell Senate Office Building door. The love on their faces filled me with new life.

At home we found the front door blocked with mounds of telegrams, packages, and flowers. We changed clothes, then got in the Corvette to drive to dinner. It wouldn't start. I realized that it was the first time I'd tried to start it since I'd been nominated, and as a result the battery was dead. I jump-started the car and we drove to Morton's of Chicago, a steak house in Tyson's Corners. The young men who did the restaurant's valet parking smiled broadly as we got out and pumped my hand vigorously. What a relief it was to deal with regular people again! The headwaiter seated us in the middle of the room, and though I briefly felt exposed and uncomfortable, the staff greeted us warmly, leaving no doubt that they were on my side. Then Bob and Mary Ellen Bork walked in, accompanied by Ted Olson, a Washington lawyer who would later become George W. Bush's solicitor general, and his fiancée, Barbara Bracher, who would later be a passenger on American Airlines flight 77 when it crashed into the Pentagon on September 11, 2001.* Seeing Judge Bork and Mary Ellen was a poignant coincidence, since they knew better than anyone else what Virginia and I were going through.

---

* This was my first meeting with Barbara. I later found out from Ricky Silberman that she had been one of my staunchest and most media-savvy supporters.

When we rose to leave at the end of the evening, the entire restaurant erupted in a spontaneous standing ovation. We also found out later that several patrons had offered to pick up our very substantial tab, but Senator Hatch had insisted on paying.

The next morning we went to church for the first time in months. After the service Virginia heard on TV that Anita had passed a polygraph examination. Sally Danforth called a few minutes later to ask whether I wanted to take one. In a phone conversation immediately after that, Larry Silberman warned me that doing so would set a precedent that might be harmful to future judicial nominees. I told them that I'd done enough to dignify the sordid proceedings and had no intention of going down that path. It was, of course, another last-ditch ploy on the part of Anita and her handlers, and nothing more would be heard of it, not even from Democratic senators who until then had seemed willing to try anything they could think of to keep me off the Court. But for the moment it caused my reviving spirits to sag yet again. Was there no limit to what my enemies would do?

A little later the White House operator patched through a call from Jehan Sadat, Anwar Sadat's widow. We had never met, and I was touched that she took the trouble to call me, though what she said touched me even more: "Judge Thomas, they are just talking about words. They are laughing at the United States around the world." I reminded her that I hadn't really said any of the things Anita had accused me of saying. "It does not matter," she repeated. "They are just words. Women around the world are suffering *real* oppression. This is *nothing* in comparison. The whole thing is silly." And so it was—except, of course, that this silliness had done great harm to my family and me. But Mrs. Sadat's own words were far from meaningless, at least to me. It buoyed me up to know that someone of her stature, who had devoted much of her life to women's issues around the world, understood what was really happening in Washington. I wasn't crazy after all.

The Judiciary Committee gave those few witnesses who were willing to testify against me virtually the whole of Sunday to do so. None was convincing. Susan Hoerchner claimed that Anita had once told her about an incident involving me, but couldn't remember when the conversation had occurred. As far as I knew, none of the hostile witnesses who were called by the committee or filed statements had worked with me when Anita did, and those who knew me personally had reason to seek revenge against me. Three former EEOC employees, for instance, claimed that I had behaved toward them in a way consistent with Anita's charges—but I had fired two of them for poor job performance and declined to reappoint the third after she'd failed to pass her bar exam. Yet all were treated with the utmost deference. It would not be until the wee hours of Monday morning that the committee was finally pressured into hearing witnesses who had a very different story to tell, including several people who had worked with me at EEOC while Anita was there and made it clear that they thought me incapable of having behaved inappropriately toward my female colleagues. Clearly the Democrats had made a concerted effort to prevent these witnesses from testifying in prime time. At one point Senator Metzenbaum went so far as to argue that there was no need to let any more of them speak on my behalf: "I wonder . . . if we couldn't stipulate that all of that testimony will be very supportive of Clarence Thomas. I don't think there's any argument about that. I don't know why there's any reason to have to hear it."

Later on I read the transcript of the hearings and saw why Senator Metzenbaum didn't care to hear from Dean Kothe, whose testimony about his dealings with Anita was devastating:

> Never once . . . in any other discussions we ever had when she was on our faculty, when she was in my home, whenever we were out together at all—at any time [did Anita Hill say] that

> Clarence Thomas was anything less than a genuinely fine person. In fact, she was very complimentary about him every time we've ever talked together. . . . I felt she was fascinated by him. She spoke of him almost as a hero. She talked of him as a devoted father. She talked to me about his untiring energy. She never ever in all of our discourse, in all those situations, ever said anything negative about him.

I didn't hear it at the time, though. Once again I had left it to Virginia to watch the hearings, though my mother and brother called at regular intervals to see how I was holding up, and Elizabeth Law telephoned throughout the day to update me on the unfolding testimony, ending each report with an energetic and encouraging exhortation to "stand firm." I went to bed before anyone had testified on my behalf, too tired to stay up any longer and too disgusted with the Judiciary Committee to care to do so. Sleep would not come, though, so after tossing and turning for a couple of hours, I got up to pray and read the Bible. At four I called Leola, who was awake and excited. She told me the hearings had ended in an uproar three hours earlier. It seemed that Senator Metzenbaum had made the mistake of tangling with John Doggett, whom I knew from Yale. After he testified about an encounter with Anita that did not show her in a good light, Senator Metzenbaum implied, on the basis of unsworn telephone interviews, that John had himself sexually harassed two women a decade earlier. Not surprisingly, John blew up. "He shut down the hearings," Leola said gleefully. "He almost came over the table!"

Virginia and I stayed home on Monday, avoiding the reporters who were camped out in front of our house and throughout the neighborhood. We spent the morning with the Laws and Jameses, reflecting on the meaning of the events of the past two weeks. We had all come to see the campaign against my confirmation as evil. There seemed no other way to explain it. My opponents could have fought my confirmation on legitimate grounds, yet they chose to tell scurrilous lies in an attempt to destroy me personally. No reason other than hatred could account for so unprecedentedly vicious an attack. Yet I knew I had to come to terms with what had happened, and ultimately to accept it: my only alternative was to let myself be driven mad by the senselessness of my ordeal, the same way Tom Robinson had been driven mad in *To Kill a Mockingbird*. When Joseph returned from the enslavement into which his brothers had sold him, he told them, "You meant it for evil, but God meant it for good." Perhaps the fires through which I had passed would have a purifying effect on me, just as a blast furnace burns the impurities out of steel. I already knew that they had brought me closer to God, and I asked Him, as I had so many times before, to help me resist the temptation to hate those who had harmed me.

On Tuesday the full Senate debated my nomination. Once again I chose not to watch the proceedings, though Senator Danforth called me in the morning to brief me. "I don't know how the vote is going to come out, Clarence, but I'll do my best for you," he said.

Later in the day he sent a messenger to the house to deliver a package. It contained the cassette of "Onward, Christian Soldiers" that he had played for Virginia and me in the bathroom of his office.

With it was a letter:

Dear Clarence—

Here is a gift to remind you of what was good about this past week. At your time of greatest weakness, you called on God for strength. He answered your prayer.

Win or lose tonight, glory to God, for He has won a great victory.

For the rest of our lives, I will always be as close as your telephone. Whatever the outcome, I will always be

Your brother in Christ,
Jack

Early that evening the Senate voted. Virginia and I invited our friends to stop by the house regardless of the outcome so that we could thank them personally for everything they'd done. So many people had done so much. I decided to take a long, hot bath to relax myself before they came.

"Do you want to listen to the roll call?" Virginia asked as I eased myself into the tub.

"Absolutely not," I said. "I don't care what they do. God never got me into anything that He didn't get me out of."

I heard the phone ring as I soaked my tired body. Virginia answered. It was Dana Barbieri, her assistant. They spoke for a moment, then she hung up and came into the bathroom. "Dana just called to congratulate you," she said. "You were confirmed. Fifty-two to forty-eight."

"Whoop-dee damn-doo," I said, sliding deeper into the comforting water. I thought of what Ray Donovan, President Reagan's much-maligned Secretary of Labor, had said after being acquitted of corruption charges: "Where do I go to get my reputation back?" Mere confirmation, even to the Supreme Court, seemed pitifully small compensation for what had been done to me.

It was raining when I came downstairs, but that didn't stop a steady stream of friends from pouring into our tiny family room. Chet Lott, Senator Trent Lott's son, the manager of a nearby Domino's Pizza outlet, generously offered to deliver what seemed like several dozen pizzas to feed the fast-growing crowd. I tried to hug and thank everybody, but it was impossible. Virginia told me that I should go outside and meet the press, but that was the last thing I wanted to do. Finally Senator Strom Thurmond talked some sense into me. "You have to go out there," he said. "You have to thank the people of America and the crowd of wonderful people out there who came to your house and stood in the rain to wish you well." It seemed ironic that a man who had once been a fervent advocate of racial segregation had ended up supporting my nomination—but, then, nearly everything about the past few months had been ironic beyond imagining. Virginia and I went outside and greeted the cheering crowd, and I said the words that I needed to hear myself say: "I think that no matter how difficult or how painful the process has been, that this is a time for healing in our country, that we have to put these things behind us."

In the weeks that followed, the flood of mail that had been coming to our house all summer became a deluge. We received letters of support, prayer offerings, invitations to use vacation homes, even McDonald's gift cards, all of them heartwarming tokens of the kindness and decency of the ordinary citizens of America. But we also received a sobering warning from the U.S. marshals who provided security for Virginia and me. They told us that it wasn't safe for us to wander about unescorted or stand in front of open windows. They also required me to wear a bulletproof vest and insisted on staying in the house around the clock for the next couple of days. Some of my opponents, the marshals explained, were now so desperate to stop me that my life might literally be in danger.

The White House hastily scheduled a public swearing-in ceremony on the South Lawn for Friday. Hundreds of family members, old friends, loyal supporters, senators, and celebrities were invited. Virginia and I drove ourselves to the White House for the ceremony. By then we were no longer under guard, which was mostly a relief but sometimes unnerving. As we waited on Duke Street to make a left turn, an eighteen-wheel tractor-trailer with New Jersey license plates pulled alongside our car and came to a complete stop, ignoring the green light. I had no idea what he was doing, and felt a momentary twinge of vulnerability. Then he rolled down his window, stuck his arm out of the cab, gave us a thumbs-up sign, and pulled away, the truck straining under a heavy load of metal pipes.

Virginia and I looked at each other in astonishment, then thanked God for the good people of this country.

We went first to St. Alban's Episcopal Church, where the Reverend Jack Danforth presided over a service of thanksgiving, delivering a moving, heartfelt homily on justice to a roomful of pews packed with the familiar faces of family and friends. Then we went to the White House, where I threw protocol to the winds and went directly to Mark Paoletta's office to thank him for his friendship and for everything he'd done to defend me. We sat among tall stacks of paper, eating hamburgers and talking over the strange events of the summer and fall. I knew how inadequate my words of gratitude sounded—it would have taken a poet to tell him how I felt—but I also knew that out of this nightmare had come a friend to whom I would remain close for the rest of my life.

It had been raining all week, but the sun pushed through the clouds and shone down on the South Lawn like a triumphant colossus. Virginia and I stood next to President and Mrs. Bush and Justice Byron White, who had agreed to administer the oath of office. (Chief Justice Rehnquist's wife had just died after a long, harrowing battle with cancer.) "America is blessed to have a man of this character serve on its highest court," the president told the crowd.

My response was simple: "There have been many difficult days as we all went through the confirmation battle. But on this sunny day in October, at the White House, there is joy. Joy in the morning." I was alluding to Psalm 30, which had brought me much comfort in recent weeks and from which a friend had suggested that I quote on this great day: *I will praise you, LORD, for you have rescued me. You refused to let my enemies triumph over me. . . . Weeping may go on all night, but joy comes with the morning.* I had forgotten that it was possible to know such joy. Thanks to God's direct intervention, I had risen phoenixlike from the ashes of self-pity and despair, and though my wounds were still raw, I trusted that in time they, too, would heal.

Afterward the president and Mrs. Bush welcomed the members of my family who had come to the ceremony. When it was all over, Leola collapsed—the accumulated stress had finally caught up with her—and the Bushes, who were the kindest of people, personally arranged an overnight hospital stay for her. The remainder of my extended family then departed for Alexandria, ending the day at my favorite Chinese restaurant, House of Dynasty. Everyone was there, including Marjorie and Don Lamp, Virginia's parents, who picked up the enormous tab; Jamal, who led us in prayer before dinner; and C, for whom I felt sadness. His involvement in my life had ended with my conception. He had missed all that had happened to me between then and now—the struggle, the heartbreak, the triumph—but he was my father, and so it was right for him to be there. I wanted him there.

IMMEDIATELY AFTER MY confirmation, I told my secretaries and law clerks that they were all to come to the Supreme Court with me. I had already hired new clerks for the coming year at the court of appeals, so I arranged for several of them to work for Mike Luttig, who had delayed his own swearing-in as an appellate judge to see me through the confirmation process. I also hired my former clerk Chris Landau, who had just finished clerking for Justice Scalia; the Supreme Court's fall term was already under way, and I knew that it would be easier to get up to speed more quickly if I brought on board some clerks with Court experience. Time was of the essence, since the November conference (the meeting at which the justices vote on cases) was less than two weeks off. Supreme Court justices normally retire at the end of a term, after which the Senate confirms their successors during the Court's summer recess, thus giving the incoming justice enough time to move his office, hire his staff, and prepare for the cases he will hear during his first term.

Justice Marshall had followed the custom and retired in June, but the protracted battle that followed his resignation had left me with little time to get ready for my first conference.

Chris called in a panic on the Monday after my swearing-in ceremony. He needed to go to work at once, but he couldn't be given access to the briefs and other essential documents until I signed the necessary personnel papers for him and the other clerks—and I couldn't do that until Chief Justice Rehnquist formally administered the judicial oath. I had wrongly supposed that the White House ceremony was sufficient. I was reluctant to disturb the Chief Justice so soon after the death of his wife, so I spoke instead to Rob Jones, his administrative assistant. Rob told me that the Chief Justice would swear me in on Wednesday if he came to the Court that day, and that I should plan to be at the Court in case he did.

I drove to the Court on Wednesday, planning to stay just long enough to be sworn in and confer with Chris. I felt at home from the moment I pulled into the underground parking area and was greeted energetically by the police officers and other Court employees. After spending a few minutes introducing myself and shaking hands all around, I went upstairs to see my new chambers for the first time. I was immediately struck by their grandeur and elegance, though the office itself was much smaller than the one I had occupied at the court of appeals; I noticed, too, that the reception area was full of flowers, packages, and bulging sacks of mail. Chris then briefed me on my long list of new duties. First of all I would need to review a dozen "holds-for-nine," the petitions for certiorari (requests for the Court to hear a case) that the Court had put on hold while waiting for the arrival of a ninth justice. After that there were twenty more cases scheduled to be heard starting on November 4. I quickly calculated that I would need to read several thousand pages of petitions, briefs, and related materials over the next two weeks, and the prospect alarmed me. Where on earth would I find the energy, strength, and concentration to do it all? What I needed was a vacation, not another marathon.

At eleven Rob Jones called to tell me that the Chief Justice was in his office and would swear me in at noon. I could invite Virginia and Senator Danforth, he said, but no one else. I called them at once and they came to my new chambers as quickly as possible. Jack took one look at the briefs lining the bookshelves and said, "This looks *really boring,* Clarence! I hope you enjoy it, but don't expect me to read many of your opinions." The three of us shared a laugh, delighted at the thought of how insignificant the discomfort of spending two weeks reading briefs would seem after what we had just endured. It felt good to be laughing again. Then we walked down quiet halls of white marble and brass to the conference room. "This looks like a mausoleum," Jack quipped. "Where are all the dead bodies?"

Chief Justice Rehnquist, who was legendary for his punctuality, came to the conference room moments after we got there, greeted us, and administered the judicial oath. The two of us signed the necessary documents. Then he shook hands with each of us, congratulated me, and left. After a month in the pillory, three months of arduous preparation, and a lifetime of near-ceaseless work, I was a member of the United States Supreme Court. Now it was time to get to work again.

The next few days were hell. Even under the best of circumstances, the sheer volume of material I had to read and absorb would have been daunting, and coming as it did on the heels of my confirmation, it was all I could do to force myself to keep going. Chief Justice Rehnquist scheduled my formal investiture for Friday, November 1, but I had no time to reflect on the occasion, for the conference would begin immediately after the ceremony, and I would sit for my first oral argument the following Monday.

Somehow I found the time to visit with my new colleagues in the week before my investiture. All of them treated me with the utmost kindness and consideration, my first indication of the way they would treat me throughout my tenure on the Court. Justice White, whose chambers I now occupy, said something in the course of our brief talk that has stayed with me ever since: "It doesn't matter how you got here. All that matters now is what you do here."

What was supposed to have been a brief courtesy call on Justice Marshall ballooned into a two-and-a-half-hour visit, and I loved every minute of it. He regaled me with tales of his long, remarkable career as a civil-rights lawyer. "I would have been shoulder to shoulder with you back then—if I'd had the courage," I said.

"I did in my time what I had to do," Justice Marshall replied. "You have to do in your time what you have to do." Those words have stayed with me, too.

The ceremony itself was simple. Virginia and I were photographed for the Court's records. Then I was seated in the "John Marshall chair" located just in front of and to the right of the long Court bench, named in honor of the longest-serving Chief Justice in the history of the Supreme Court. The courtroom was packed with friends, family, Court officials, and members of the Supreme Court Bar. Acting Attorney General William P. Barr moved that the Clerk of the Court read the presidential commission issued to me as an associate justice. Chief Justice Rehnquist granted the motion, and the clerk read:

> Know Ye; That reposing special trust and confidence in the Wisdom, Uprightness, and Learning of Clarence Thomas, of Georgia, I have nominated, and, by and with the advice and consent of the Senate, do appoint him an Associate Justice of the United States and do authorize and empower him to execute and fulfill the duties of that Office according to the Constitution and Laws of the said United States. . . .

I heard the words, but the man who was reading them was a million miles away. Without warning, memories of home, my grandparents, and the accumulated toil of the last four decades swirled through my mind, numbing me to the proceedings. I thought I was going to collapse as pent-up emotions erupted within me. I looked at Virginia, always loving but hurt by the confirmation process and anxious about the future; I looked at Leola, stripped of thirty pounds by worry and stress. Above all I thought of my grandparents, wishing more than anything else that they could be with me. They had believed, worked, and sacrificed, not knowing where their labors would lead but trusting that their boys would do well. They could not possibly have dreamed of this moment—yet it was theirs.

I heard the distant, authoritative voice of the Chief Justice of the United States of America: "I now ask the chief deputy clerk of the Court to escort Justice Thomas to the bench." I walked with suddenly stiff legs to the steps of the raised bench and saw the seven men and one woman with whom I would now sit. They looked at me, imposing but pleasant. My mouth went dry and my tongue grew thick.

"Justice Thomas, are you prepared to take the oath?"

"I am."

"Then repeat after me: I, Clarence Thomas, do solemnly swear that I will administer justice without respect to person, and do equal right to the poor and to the rich, and that I will faithfully and impartially discharge and perform all of the duties incumbent upon me as Associate Justice of the Supreme Court of the United States under the Constitution and laws of the United States. So help me God."

Struggling to control my surging emotions, I repeated the oath, thinking as I did so of how Daddy and Aunt Tina had raised me to fulfill it. *Any job worth doing,* they had told me, *is worth doing right.* This, I knew, was a job worth doing.

"Justice Thomas, on behalf of all the members of this court, it is a pleasure to extend to you a very warm welcome as an associate justice of this Court and to wish for you a long and happy career in our common calling," the Chief Justice said. I felt one last twinge of panic. Would I have to say something? Could I? My mouth seemed not to want to work. But no: all I had to do was shake hands with my new colleagues.

I stayed only briefly at the reception, for there was no time to linger. The traditional walk with the Chief Justice down the front steps of the Court had been scheduled to follow the investiture, after which my first conference would begin immediately. I greeted as many people as I could in the few moments I could spare. Just before I was whisked away, a black man came up to me and introduced himself as Gary Kemp, a newly hired deputy clerk of the Court. "I have copies of your birth record," he told me. I flinched. Who was this man, and why did he have such a personal document? Having stood by for months as my life was plundered and twisted, I reflexively assumed that something bad was about to happen, the same way a battered child might draw back from the touch of a kindly stranger.

"I am the great-grandson of Mrs. Lula Kemp," he said.

"Who was she?" Then it came back to me: she was the midwife from Sandfly who had come to Pinpoint in 1948 to bring me into the world. I broke into a broad smile. How amazing—and how fitting, even providential—that the great-grandson of the woman who had attended my birth was present at the end of my last difficult journey, just as she had been present at the end of my first.

I longed to talk to Gary about Pinpoint and Sandfly and the lost world of my youth . . . but there was no more time. Chief Justice Rehnquist was ready to go. I walked with him into the awe-inspiring Great Hall of the Court and through the imposing front doorway, glancing at the huge brass doors. As we walked slowly down the gleaming white marble steps, lit by the brightness of a beautiful sunny morning, I thought back to another sunny day in 1955, the day when my brother and I had walked to the white house at 542 East Thirty-second Street to live with our grandparents, all of our belongings stuffed into a pair of grocery bags. So began the journey that had led me at last to these steps. It was there that Daddy and Aunt Tina taught me all they knew and gave me all they had: their wisdom, their energy, their way of life. For a long while I had, like the prodigal son, abandoned them and what they had taught me. I had come back too late to reconcile with them in this life, but at least I had returned to their ways, vowing to live my life as a memorial to theirs. Mere weeks before I had looked on in horror and disgust as my memorial and their legacy were distorted, caricatured, and painted over with the graffiti of cheap, tawdry lies. The life I had pledged to live in their memory had been desecrated. Now I had to start building another monument to them and all they stood for—one grain of sand at a time.

I felt their presence as the Chief Justice escorted me into the elegant conference room. Looking at my new colleagues as the heavy door closed firmly behind me, I thanked God for the lives of the man and woman who had led me here.

Then I prayed silently:

Lord, grant me the wisdom to know what is right and the courage to do it. Amen.